Earthquake Engineering

Earthquake Engineering

Edited by
Devin Burns

Larsen & Keller
www.larsen-keller.com

Earthquake Engineering
Edited by Devin Burns
ISBN: 978-1-63549-090-9 (Hardback)

▤ Larsen & Keller

Published by Larsen and Keller Education,
5 Penn Plaza,
19th Floor,
New York, NY 10001, USA

Cataloging-in-Publication Data

Earthquake engineering / edited by Devin Burns.
   p. cm.
Includes bibliographical references and index.
ISBN 978-1-63549-090-9
1. Earthquake engineering. 2. Earthquake resistant design. 3. Seismology.
I. Burns, Devin.
TA654.6 .E27 2017
624.176 2--dc23

The publisher's policy is to use permanent paper from mills that operate a sustainable forestry policy. Furthermore, the publisher ensures that the text paper and cover boards used have met acceptable environmental accreditation standards.

Printed and bound in the United States of America.

For more information regarding Larsen and Keller Education and its products, please visit the publisher's website www.larsen-keller.com

Table of Contents

Preface

Earthquake engineering is the study of design of buildings that are resistant to earthquakes. Earthquake engineering is an interdisciplinary branch and encompasses disciplines ranging from civil engineering to geotechnical engineering. Topics included in this text seek to provide comprehensive information about earthquake-resistant engineering and effective earthquake exposure methods. For someone with an interest and eye for detail, this book covers the most significant topics in the field of earthquake engineering. This textbook seeks to shed light on some of the novel concepts of earthquake engineering. It will prove vital for students studying in the fields of civil engineering, seismology and structural engineering.

A detailed account of the significant topics covered in this book is provided below:

Chapter 1- Earthquake engineering is a branch of engineering that deals with the design and construction of structures like buildings, bridges etc. in such a manner that they are able to withstand seismic activity. These structures are also capable against other natural calamities. This introduction provides an overview of the interdisciplinary field of earthquake engineering.

Chapter 2- Earthquakes are a result of seismic activity in the Earth's crust. In this chapter the reader is informed about the causes of earthquakes and its various types such as megathrust earthquake, intraplate earthquake, interplate earthquake, deep-Focus earthquake and volcano tectonic earthquake. The features of each type of earthquake have been discussed to enable an insightful understanding of the subject of earthquakes.

Chapter 3- To measure seismic activity in the Earth's crust, a multitude of scales have been developed and each of these scales is based on measuring components like the energy released and the after-effects of the earthquake. The chapter studies the moment magnitude scale, the Richter scale, Mercalli intensity scale and the Environmental Seismic Intensity scale. There is a section on the earthquake shaking table which is a device that shakes structural models and building components in a vast range of stimulated motions, helping predict how structures react to earthquakes.

Chapter 4- The study of seismic events, their causes, effects and prediction is known as seismology. This chapter provides comprehensive information on seismology, paleoseismology and seismometers and seismic waves. The content elaborately explores the fields and lists the types, methodology and terminology related to each of these topics.

Chapter 5- Seismic activity can cause an insurmountable amount of damage to human lives, cities and the environment as well. To protect lives and man-made structures has been the main focus of earthquake engineering. Three major techniques to achieve this have been discussed in this chapter- vibration control, vibration isolation and seismic retrofit. These techniques are applied by various devices that are fitted into buildings to mitigate the damage of earthquakes.

Chapter 6- Earthquake resistant structures are built to withstand earthquakes and to minimize damage to life and property. Ancient architecture worked around this problem by constructing stiff and strong structures but with the modern advancements in architectural science, the focus has shifted to keeping functionality intact while also incorporating elements that will help structures suffer less damage. To this end the chapter explores the inclusion of steel plate shear walls, its advantages and analytical models.

Chapter 7- Earthquake engineering requires input from other fields to develop and design structures that can withstand earthquakes. One of these branches, known as structural engineering is concerned with the design of structures, creative use of building materials and economy in construction. Geotechnical engineering on the other hand focuses on the soil conditions and site investigations which delve into aspects like fault distribution, soil erosion, subsurface categorization, bedrock properties etc. This chapter inspects the allied fields of geotechnical engineering, civil engineering, seismic analysis and structural engineering.

Chapter 8- Disaster management or emergency management closely considers contingency planning and decreasing the damage caused by disasters. It seeks to provide information as well as material for effective responses to disasters. This chapter studies peak ground acceleration and emergency management in an effort to comprehensively study the essential aspects of earthquake engineering.

I would like to make a special mention of my publisher who considered me worthy of this opportunity and also supported me throughout the process. I would also like to thank the editing team at the back-end who extended their help whenever required.

Editor

Introduction to Earthquake Engineering

Earthquake engineering is a branch of engineering that deals with the design and construction of structures like buildings, bridges etc. in such a manner that they are able to withstand seismic activity. These structures are also capable against other natural calamities. This introduction provides an overview of the interdisciplinary field of earthquake engineering.

Earthquake engineering is an interdisciplinary branch of engineering that designs and analyzes structures, such as buildings and bridges, with earthquakes in mind. Its overall goal is to make such structures more resistant to earthquakes. An earthquake (or seismic) engineer aims to construct structures that will not be damaged in minor shaking and will avoid serious damage or collapse in a major earthquake. Earthquake engineering is the scientific field concerned with protecting society, the natural environment, and the man-made environment from earthquakes by limiting the seismic risk to socio-economically acceptable levels. Traditionally, it has been narrowly defined as the study of the behavior of structures and geo-structures subject to seismic loading; it is considered as a subset of structural engineering, geotechnical engineering, mechanical engineering, chemical engineering, applied physics, etc. However, the tremendous costs experienced in recent earthquakes have led to an expansion of its scope to encompass disciplines from the wider field of civil engineering, mechanical engineering and from the social sciences, especially sociology, political science, economics and finance.

Shake-table crash testing of a regular building model (left) and a base-isolated building model (right) at UCSD

The main objectives of earthquake engineering are:

- Foresee the potential consequences of strong earthquakes on urban areas and civil infrastructure.

- Design, construct and maintain structures to perform at earthquake exposure up to the expectations and in compliance with building codes.

A properly engineered structure does not necessarily have to be extremely strong or expensive. It has to be properly designed to withstand the seismic effects while sustaining an acceptable level of damage.

Seismic Loading

Tokyo Skytree, equipped with a tuned mass damper, is the world's
tallest tower and is the world's second tallest structure.

Seismic loading means application of an earthquake-generated excitation on a structure (or geo-structure). It happens at contact surfaces of a structure either with the ground, with adjacent structures, or with gravity waves from tsunami. The loading that is expected at a given location on the Earth's surface is estimated by engineering seismology. It is related to the seismic hazard of the location.

Seismic Performance

Earthquake or seismic performance defines a structure's ability to sustain its main functions, such as its safety and serviceability, *at* and *after* a particular earthquake exposure. A structure is normally considered *safe* if it does not endanger the lives and well-being of those in or around it by partially or completely collapsing. A structure may be considered *serviceable* if it is able to fulfill its operational functions for which it was designed.

Basic concepts of the earthquake engineering, implemented in the major building codes, assume that a building should survive a rare, very severe earthquake by sustaining significant damage but without globally collapsing. On the other hand, it should remain operational for more frequent, but less severe seismic events.

Seismic Performance Assessment

Engineers need to know the quantified level of the actual or anticipated seismic performance associated with the direct damage to an individual building subject to a specified ground shaking. Such an assessment may be performed either experimentally or analytically.

Experimental Assessment

Experimental evaluations are expensive tests that are typically done by placing a (scaled) model of the structure on a shake-table that simulates the earth shaking and observing its behavior. Such kinds of experiments were first performed more than a century ago. Only recently has it become possible to perform 1:1 scale testing on full structures.

Due to the costly nature of such tests, they tend to be used mainly for understanding the seismic behavior of structures, validating models and verifying analysis methods. Thus, once properly validated, computational models and numerical procedures tend to carry the major burden for the seismic performance assessment of structures.

Analytical/Numerical Assessment

Snapshot from shake-table video of a 6-story non-ductile concrete building destructive testing

Seismic performance assessment or seismic structural analysis is a powerful tool of earthquake engineering which utilizes detailed modelling of the structure together with methods of structural analysis to gain a better understanding of seismic performance of building and non-building structures. The technique as a formal concept is a relatively recent development.

In general, seismic structural analysis is based on the methods of structural dynamics. For decades, the most prominent instrument of seismic analysis has been the earthquake response spectrum method which also contributed to the proposed building code's concept of today.

However, such methods are good only for linear elastic systems, being largely unable to model the structural behavior when damage (i.e., non-linearity) appears. Numerical *step-by-step integration* proved to be a more effective method of analysis for multi-degree-of-freedom structural systems with significant non-linearity under a transient process of ground motion excitation.

Basically, numerical analysis is conducted in order to evaluate the seismic performance of buildings. Performance evaluations are generally carried out by using nonlinear static pushover analysis or nonlinear time-history analysis. In such analyses, it is essential to achieve accurate non-linear modeling of structural components such as beams, columns, beam-column joints, shear walls etc. Thus, experimental results play an important role in determining the modeling parameters of individual components, especially those that are subject to significant non-linear deformations. The individual components are then assembled to create a full non-linear model of the structure. Thus created models are analyzed to evaluate the performance of buildings.

The capabilities of the structural analysis software are a major consideration in the above process as they restrict the possible component models, the analysis methods available and, most importantly, the numerical robustness. The latter becomes a major consideration for structures that venture into the non-linear range and approach global or local collapse as the numerical solution becomes increasingly unstable and thus difficult to reach. There are several commercially available Finite Element Analysis software's such as CSI-SAP2000 and CSI-PERFORM-3D and Scia Engineer-ECtools which can be used for the seismic performance evaluation of buildings. Moreover, there is research-based finite element analysis platforms such as OpenSees, RUAUMOKO and the older DRAIN-2D/3D, several of which are now open source.

Research for Earthquake Engineering

Research for earthquake engineering means both field and analytical investigation or experimentation intended for discovery and scientific explanation of earthquake engineering related facts, revision of conventional concepts in the light of new findings, and practical application of the developed theories.

Shake-table testing of Friction Pendulum Bearings at EERC

The National Science Foundation (NSF) is the main United States government agency that supports fundamental research and education in all fields of earthquake engineering. In particular, it focuses on experimental, analytical and computational research on design and performance enhancement of structural systems.

E-Defense Shake Table

The Earthquake Engineering Research Institute (EERI) is a leader in dissemination of earthquake engineering research related information both in the U.S. and globally.

A definitive list of earthquake engineering research related shaking tables around the world may be found in Experimental Facilities for Earthquake Engineering Simulation Worldwide. The most prominent of them is now E-Defense Shake Table in Japan.

Major U.S. Research Programs

NSF also supports the George E. Brown, Jr. Network for Earthquake Engineering Simulation.

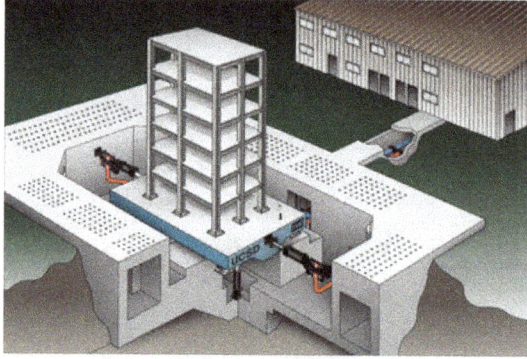

Large High Performance Outdoor Shake Table, UCSD, NEES network

The NSF Hazard Mitigation and Structural Engineering program (HMSE) supports research on new technologies for improving the behavior and response of structural systems subject to earthquake hazards; fundamental research on safety and reliability of constructed systems; innovative developments in analysis and model based simulation of structural behavior and response including soil-structure interaction; design concepts that improve structure performance and flexibility; and application of new control techniques for structural systems.

(NEES) that advances knowledge discovery and innovation for earthquakes and tsunami loss reduction of the nation's civil infrastructure and new experimental simulation techniques and instrumentation.

The NEES network features 14 geographically-distributed, shared-use laboratories that support several types of experimental work: geotechnical centrifuge research, shake-table tests, large-scale structural testing, tsunami wave basin experiments, and field site research. Participating universities include: Cornell University; Lehigh University; Oregon State University; Rensselaer Polytechnic Institute; University at Buffalo, State University of New York; University of California, Berkeley; University of California, Davis; University of California, Los Angeles; University of California, San Diego; University of California, Santa Barbara; University of Illinois, Urbana-Champaign; University of Minnesota; University of Nevada, Reno; and the University of Texas, Austin.

NEES at Buffalo testing facility

The equipment sites (labs) and a central data repository are connected to the global earthquake engineering community via the NEEShub website. The NEES website is powered by HUBzero software developed at Purdue University for nanoHUB specifically to help the scientific community share resources and collaborate. The cyberinfrastructure, connected via Internet2, provides interactive simulation tools, a simulation tool development area, a curated central data repository, animated presentations, user support, telepresence, mechanism for uploading and sharing resources, and statistics about users and usage patterns.

This cyberinfrastructure allows researchers to: securely store, organize and share data within a standardized framework in a central location; remotely observe and participate in experiments through the use of synchronized real-time data and video; collaborate with colleagues to facilitate the planning, performance, analysis, and publication of research experiments; and conduct computational and hybrid simulations that may combine the results of multiple distributed experiments and link physical experiments with computer simulations to enable the investigation of overall system performance.

These resources jointly provide the means for collaboration and discovery to improve the seismic design and performance of civil and mechanical infrastructure systems.

Earthquake Simulation

The very first earthquake simulations were performed by statically applying some *horizontal inertia forces* based on scaled peak ground accelerations to a mathematical model of a building. With the further development of computational technologies, static approaches began to give way to dynamic ones.

Dynamic experiments on building and non-building structures may be physical, like shake-table testing, or virtual ones. In both cases, to verify a structure's expected seismic performance, some researchers prefer to deal with so called "real time-histories" though the last cannot be "real" for a hypothetical earthquake specified by either a building code or by some particular research requirements. Therefore, there is a strong incentive to engage an earthquake simulation which is the seismic input that possesses only essential features of a real event.

Sometimes earthquake simulation is understood as a re-creation of local effects of a strong earth shaking.

Structure Simulation

Theoretical or experimental evaluation of anticipated seismic performance mostly requires a structure simulation which is based on the concept of structural likeness or similarity. Similarity is some degree of analogy or resemblance between two or more objects. The notion of similarity rests either on exact or approximate repetitions of patterns in the compared items.

In general, a building model is said to have similarity with the real object if the two share *geometric similarity*, *kinematic similarity* and *dynamic similarity*. The most vivid and effective type of similarity is the *kinematic* one. *Kinematic similarity* exists when the paths and velocities of moving particles of a model and its prototype are similar.

Concurrent experiments with two building models which are *kinematically equivalent* to a real prototype.

The ultimate level of *kinematic similarity* is *kinematic equivalence* when, in the case of earthquake engineering, time-histories of each story lateral displacements of the model and its prototype would be the same.

Seismic Vibration Control

Seismic vibration control is a set of technical means aimed to mitigate seismic impacts in building and non-building structures. All seismic vibration control devices may be classified as *passive*, *active* or *hybrid* where:

- *passive control devices* have no feedback capability between them, structural elements and the ground;

- *active control devices* incorporate real-time recording instrumentation on the ground integrated with earthquake input processing equipment and actuators within the structure;

- *hybrid control devices* have combined features of active and passive control systems.

When ground seismic waves reach up and start to penetrate a base of a building, their energy flow density, due to reflections, reduces dramatically: usually, up to 90%. However, the remaining portions of the incident waves during a major earthquake still bear a huge devastating potential.

After the seismic waves enter a superstructure, there are a number of ways to control them in order to soothe their damaging effect and improve the building's seismic performance, for instance:

- to dissipate the wave energy inside a superstructure with properly engineered dampers;

- to disperse the wave energy between a wider range of frequencies;

- to absorb the resonant portions of the whole wave frequencies band with the help of so-called *mass dampers*.

Mausoleum of Cyrus, the oldest base-isolated structure in the world

Devices of the last kind, abbreviated correspondingly as TMD for the tuned (*passive*), as AMD for the *active*, and as HMD for the *hybrid mass dampers*, have been studied and installed in high-rise buildings, predominantly in Japan, for a quarter of a century.

However, there is quite another approach: partial suppression of the seismic energy flow into the superstructure known as seismic or base isolation.

For this, some pads are inserted into or under all major load-carrying elements in the base of the building which should substantially decouple a superstructure from its substructure resting on a shaking ground.

The first evidence of earthquake protection by using the principle of base isolation was discovered in Pasargadae, a city in ancient Persia, now Iran, and dates back to the 6th century BCE. Below, there are some samples of seismic vibration control technologies of today.

Dry-stone Walls Control

People of Inca civilization were masters of the polished '*dry-stone walls*', called ashlar, where blocks of stone were cut to fit together tightly without any mortar. The Incas were among the best stonemasons the world has ever seen and many junctions in their masonry were so perfect that even blades of grass could not fit between the stones.

Dry-stone walls of Machu Picchu Temple of the Sun, Peru

Peru is a highly seismic land and for centuries the mortar-free construction proved to be apparently more earthquake-resistant than using mortar. The stones of the dry-stone walls built by the Incas could move slightly and resettle without the walls collapsing, a passive structural control technique employing both the principle of energy dissipation and that of suppressing resonant amplifications.

Lead Rubber Bearing

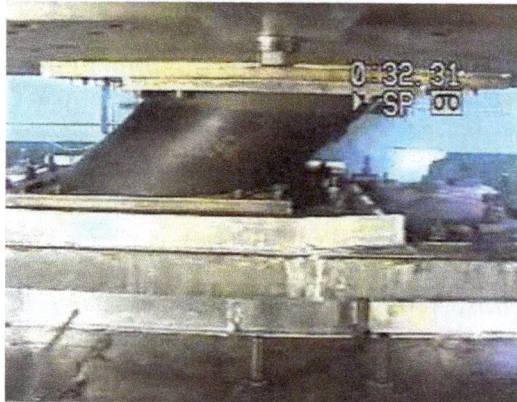
LRB being tested at the UCSD Caltrans-SRMD facility

Lead Rubber Bearing or LRB is a type of base isolation employing a heavy damping. It was invented by Bill Robinson, a New Zealander.

Heavy damping mechanism incorporated in vibration control technologies and, particularly, in base isolation devices, is often considered a valuable source of suppressing vibrations thus enhancing a building's seismic performance. However, for the rather pliant systems such as base isolated structures, with a relatively low bearing stiffness but with a high damping, the so-called "damping force" may turn out the main pushing force at a strong earthquake. The video shows a Lead Rubber Bearing being tested at the UCSD Caltrans-SRMD facility. The bearing is made of rubber with a lead core. It was a uniaxial test in which the bearing was also under a full structure load. Many buildings and bridges, both in New Zealand and elsewhere, are protected with lead dampers and lead and rubber bearings. Te Papa Tongarewa, the national museum of New Zealand, and the New Zealand Parliament Buildings have been fitted with the bearings. Both are in Wellington which sits on an active earthquake fault.

Tuned Mass Damper

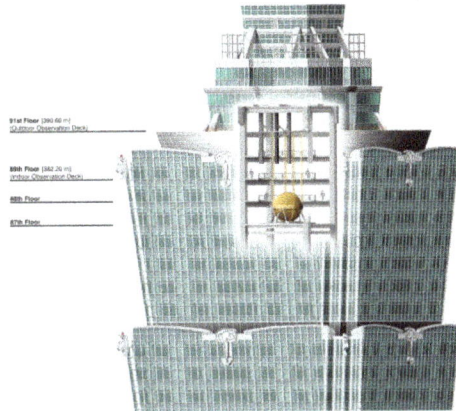

Tuned mass damper in Taipei 101, the world's third tallest skyscraper

Typically the tuned mass dampers are huge concrete blocks mounted in skyscrapers or other structures and moved in opposition to the resonance frequency oscillations of the structures by means of some sort of spring mechanism.

Taipei 101 skyscraper needs to withstand typhoon winds and earthquake tremors common in its area of the Asia-Pacific. For this purpose, a steel pendulum weighing 660 metric tones that serves as a tuned mass damper was designed and installed atop the structure. Suspended from the 92nd to the 88th floor, the pendulum sways to decrease resonant amplifications of lateral displacements in the building caused by earthquakes and strong gusts.

Friction Pendulum Bearing

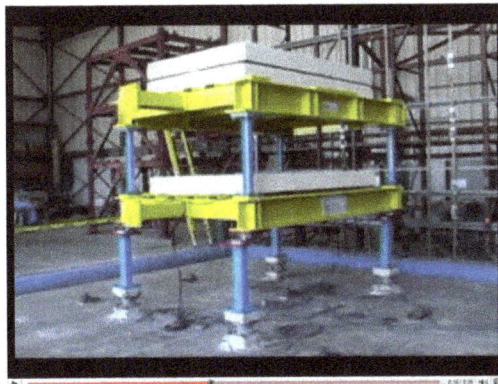

FPB shake-table testing

Friction Pendulum Bearing (FPB) is another name of Friction Pendulum System (FPS). It is based on three pillars:

- articulated friction slider;

- spherical concave sliding surface;

- enclosing cylinder for lateral displacement restraint.

Snapshot with the link to video clip of a shake-table testing of FPB system supporting a rigid building model is presented at the right.

Building Elevation Control

Transamerica Pyramid building

Building elevation control is a valuable source of vibration control of seismic loading. Pyramid-shaped skyscrapers continue to attract the attention of architects and engineers because such structures promise a better stability against earthquakes and winds. The elevation configuration can prevent buildings' resonant amplifications because a properly configured building disperses the shear wave energy between a wide range of frequencies.

Earthquake or wind quieting ability of the elevation configuration is provided by a specific pattern of multiple reflections and transmissions of vertically propagating waves, which are generated by breakdowns into homogeneity of story layers, and a taper. Any abrupt changes of the propagating waves velocity result in a considerable dispersion of the wave energy between a wide ranges of frequencies thus preventing the resonant displacement amplifications in the building.

A tapered profile of a building is not a compulsory feature of this method of structural control. A similar resonance preventing effect can be also obtained by a proper *tapering* of other characteristics of a building structure, namely, its mass and stiffness. As a result, the building elevation configuration techniques permit an architectural design that may be both attractive and functional.

Simple Roller Bearing

Simple roller bearing is a base isolation device which is intended for protection of various building and non-building structures against potentially damaging lateral impacts of strong earthquakes.

This metallic bearing support may be adapted, with certain precautions, as a seismic isolator to skyscrapers and buildings on soft ground. Recently, it has been employed under the name of *Metallic Roller Bearing* for a housing complex (17 stories) in Tokyo, Japan.

Springs-with-damper Base Isolator

Springs-with-damper close-up

Springs-with-damper base isolator installed under a three-story town-house, Santa Monica, California is shown on the photo taken prior to the 1994 Northridge earthquake exposure. It is a base isolation device conceptually similar to *Lead Rubber Bearing*.

One of two three-story town-houses like this, which was well instrumented for recording of both vertical and horizontal accelerations on its floors and the ground, has survived a severe shaking during the Northridge earthquake and left valuable recorded information for further study.

Hysteretic Damper

Hysteretic damper is intended to provide better and more reliable seismic performance than that of a conventional structure at the expense of the seismic input energy dissipation. There are five major groups of hysteretic dampers used for the purpose, namely:

- Fluid viscous dampers (FVDs)
- Metallic yielding dampers (MYDs)
- Viscoelastic dampers (VEDs)
- Friction dampers (FDs)
- Straddlingpendulum dampers (swing)

Each group of dampers has specific characteristics, advantages and disadvantages for structural applications.

Seismic Design

Seismic design is based on authorized engineering procedures, principles and criteria meant to design or retrofit structures subject to earthquake exposure. Those criteria are only consistent with the contemporary state of the knowledge about earthquake engineering structures. Therefore, a building design which exactly follows seismic code regulations does not guarantee safety against collapse or serious damage.

The price of poor seismic design may be enormous. Nevertheless, seismic design has always been a trial and error process whether it was based on physical laws or on empirical knowledge of the structural performance of different shapes and materials.

San Francisco City Hall destroyed by 1906 earthquake and fire.

San Francisco after the 1906 earthquake and fire

To practice seismic design, seismic analysis or seismic evaluation of new and existing civil engineering projects, an engineer should, normally, pass examination on *Seismic Principles* which, in the State of California, include:

- Seismic Data and Seismic Design Criteria

- Seismic Characteristics of Engineered Systems

- Seismic Forces

- Seismic Analysis Procedures

- Seismic Detailing and Construction Quality Control

To build up complex structural systems, seismic design largely uses the same relatively small number of basic structural elements (to say nothing of vibration control devices) as any non-seismic design project.

Normally, according to building codes, structures are designed to "withstand" the largest earthquake of a certain probability that is likely to occur at their location. This means the loss of life should be minimized by preventing collapse of the buildings.

Seismic design is carried out by understanding the possible failure modes of a structure and providing the structure with appropriate strength, stiffness, ductility, and configuration to ensure those modes cannot occur.

Seismic Design Requirements

Seismic design requirements depend on the type of the structure, locality of the project and its authorities which stipulate applicable seismic design codes and criteria. For instance, California Department of Transportation's requirements called *The Seismic Design Criteria* (SDC) and aimed at the design of new bridges in California incorporate an innovative seismic performance-based approach.

The most significant feature in the SDC design philosophy is a shift from a *force-based assessment* of seismic demand to a *displacement-based assessment* of demand and capacity. Thus, the newly adopted displacement approach is based on comparing the *elastic displacement* demand to the *inelastic displacement* capacity of the primary structural components while ensuring a minimum level of inelastic capacity at all potential plastic hinge locations.

The Metsamor Nuclear Power Plant was closed after the 1988 Armenian earthquake

In addition to the designed structure itself, seismic design requirements may include a *ground stabilization* underneath the structure: sometimes, heavily shaken ground breaks up which leads to collapse of the structure sitting upon it. The following topics should be of primary concerns: liquefaction; dynamic lateral earth pressures on retaining walls; seismic slope stability; earthquake-induced settlement.

Nuclear facilities should not jeopardise their safety in case of earthquakes or other hostile external events. Therefore, their seismic design is based on criteria far more stringent than those applying to non-nuclear facilities. The Fukushima I nuclear accidents and damage to other nuclear facilities that followed the 2011 Tōhoku earthquake and tsunami have, however, drawn attention to ongoing concerns over Japanese nuclear seismic design standards and caused other many governments to re-evaluate their nuclear programs. Doubt has also been expressed over the seismic evaluation and design of certain other plants, including the Fessenheim Nuclear Power Plant in France.

Failure Modes

Failure mode is the manner by which an earthquake induced failure is observed. It, generally, describes the way the failure occurs. Though costly and time consuming, learning from each real earthquake failure remains a routine recipe for advancement in *seismic design* methods. Below, some typical modes of earthquake-generated failures are presented.

The lack of reinforcement coupled with poor mortar and inadequate roof-to-wall ties can result in substantial damage to an unreinforced masonry building. Severely cracked or leaning walls are some of the most common earthquake damage. Also hazardous is the damage that may occur between the walls and roof or floor diaphragms. Separation between the framing and the walls can jeopardize the vertical support of roof and floor systems.

Soft story effect. Absence of adequate shear walls on the ground level caused damage to this structure. A close examination of the image reveals that the rough board siding, once covered by a brick veneer, has been completely dismantled from the studwall. Only the rigidity of the floor above combined with the support on the two hidden sides by continuous walls, not penetrated with large doors as on the street sides, is preventing full collapse of the structure.

Effects of soil liquefaction during the 1964 Niigata earthquake

Soil liquefaction. In the cases where the soil consists of loose granular deposited materials with the tendency to develop excessive hydrostatic pore water pressure of sufficient magnitude and compact, liquefaction of those loose saturated deposits may result in non-uniform settlements and tilting of structures. This caused major damage to thousands of buildings in Niigata, Japan during the 1964 earthquake.

Car smashed by landslide rock, 2008 Sichuan earthquake

Landslide rock fall. A landslide is a geological phenomenon which includes a wide range of ground movement, including rock falls. Typically, the action of gravity is the primary driving force for a landslide to occur though in this case there was another contributing factor which affected the original slope stability: the landslide required an *earthquake trigger* before being released.

Effects of pounding against adjacent building, Loma Prieta

Pounding against adjacent building. This is a photograph of the collapsed five-story tower, St. Joseph's Seminary, Los Altos, California which resulted in one fatality. During Loma Prieta earthquake, the tower pounded against the independently vibrating adjacent building behind. A possibility of pounding depends on both buildings' lateral displacements which should be accurately estimated and accounted for.

Effects of completely shattered joints of concrete frame, Northridge

At Northridge earthquake, the Kaiser Permanente concrete frame office building had joints completely shattered, revealing inadequate confinement steel, which resulted in the second story collapse. In the transverse direction, composite end shear walls, consisting of two wythes of brick and a layer of shotcrete that carried the lateral load, peeled apart because of inadequate through-ties and failed.

7-story reinforced concrete buildings on steep slope collapse due to the following:

- Improper construction site on a foothill.

- Poor detailing of the reinforcement (lack of concrete confinement in the columns and at the beam-column joints, inadequate splice length).

- Seismically weak soft story at the first floor.

- Long cantilevers with heavy dead load.

Sliding off foundations effect of a relatively rigid residential building structure during 1987 Whittier Narrows earthquake. The magnitude 5.9 earthquake pounded the Garvey West Apartment building in Monterey Park, California and shifted its superstructure about 10 inches to the east on its foundation.

If a superstructure is not mounted on a base isolation system, its shifting on the basement should be prevented.

Insufficient shear reinforcement let main rebars to buckle, Northridge

Reinforced concrete column burst at Northridge earthquake due to insufficient shear reinforcement mode which allows main reinforcement to buckle outwards. The deck unseated at the hinge and failed in shear. As a result, the La Cienega-Venice underpass section of the 10 Freeway collapsed.

Support-columns and upper deck failure, Loma Prieta earthquake

Loma Prieta earthquake: side view of reinforced concrete support-columns failure which triggered the upper deck collapse onto the lower deck of the two-level Cypress viaduct of Interstate Highway 880, Oakland, CA.

Failure of retaining wall due to ground movement, Loma Prieta

Retaining wall failure at Loma Prieta earthquake in Santa Cruz Mountains area: prominent northwest-trending extensional cracks up to 12 cm (4.7 in) wide in the concrete spillway to Austrian Dam, the north abutment.

Lateral spreading mode of ground failure, Loma Prieta

Ground shaking triggered soil liquefaction in a subsurface layer of sand, producing differential lateral and vertical movement in an overlying carapace of unliquified sand and silt. This mode of ground failure, termed lateral spreading, is a principal cause of liquefaction-related earthquake damage.

Severely damaged building of Agriculture Development Bank of China after 2008 Sichuan earthquake: most of the beams and pier columns are sheared. Large diagonal cracks in masonry and veneer are due to in-plane loads while abrupt settlement of the right end of the building should be attributed to a landfill which may be hazardous even without any earthquake.

Tsunami strikes Ao Nang,

Twofold tsunami impact: sea waves hydraulic pressure and inundation. Thus, the Indian Ocean earthquake of December 26, 2004, with the epicenter off the west coast of Sumatra, Indonesia, triggered a series of devastating tsunamis, killing more than 230,000 people in eleven countries by inundating surrounding coastal communities with huge waves up to 30 meters (100 feet) high.

Earthquake-resistant Construction

Earthquake construction means implementation of seismic design to enable building and non-building structures to live through the anticipated earthquake exposure up to the expectations and in compliance with the applicable building codes.

Construction of Pearl River Tower X-bracing to resist lateral forces of earthquakes and winds

Design and construction are intimately related. To achieve a good workmanship, detailing of the members and their connections should be as simple as possible. As any construction in general, earthquake construction is a process that consists of the building, retrofitting or assembling of infrastructure given the construction materials available.

The destabilizing action of an earthquake on constructions may be *direct* (seismic motion of the ground) or *indirect* (earthquake-induced landslides, soil liquefaction and waves of tsunami).

A structure might have all the appearances of stability, yet offer nothing but danger when an earthquake occurs. The crucial fact is that, for safety, earthquake-resistant construction techniques are as important as quality control and using correct materials. *Earthquake contractor* should be registered in the state of the project location, bonded and insured.

To minimize possible losses, construction process should be organized with keeping in mind that earthquake may strike any time prior to the end of construction.

Each construction project requires a qualified team of professionals who understand the basic features of seismic performance of different structures as well as construction management.

Adobe Structures

Around thirty percent of the world's population lives or works in earth-made construction. Adobe type of mud bricks is one of the oldest and most widely used building materials. The use of adobe is very common in some of the world's most hazard-prone regions, traditionally across Latin America, Africa, Indian subcontinent and other parts of Asia, Middle East and Southern Europe.

Adobe buildings are considered very vulnerable at strong quakes. However, multiple ways of seismic strengthening of new and existing adobe buildings are available.

Key factors for the improved seismic performance of adobe construction are:

- Quality of construction.

- Compact, box-type layout.

- Seismic reinforcement.

Limestone and Sandstone Structures

Base-isolated City and County Building, Salt Lake City, Utah

Limestone is very common in architecture, especially in North America and Europe. Many landmarks across the world are made of limestone. Many medieval churches and castles in Europe are made of limestone and sandstone masonry. They are the long-lasting materials but their rather heavy weight is not beneficial for adequate seismic performance.

Application of modern technology to seismic retrofitting can enhance the survivability of unreinforced masonry structures. As an example, from 1973 to 1989, the Salt Lake City and County Building in Utah was exhaustively renovated and repaired with an emphasis on preserving historical accuracy in appearance. This was done in concert with a seismic upgrade that placed the weak sandstone structure on base isolation foundation to better protect it from earthquake damage.

Timber Frame Structures

Anne Hvide's House, Denmark (1560)

Timber framing dates back thousands of years, and has been used in many parts of the world during various periods such as ancient Japan, Europe and medieval England in localities where timber was in good supply and building stone and the skills to work it were not.

The use of timber framing in buildings provides their complete skeletal framing which offers some structural benefits as the timber frame, if properly engineered, lends itself to better *seismic survivability*.

Light-frame Structures

A two-story wooden-frame for a residential building structure

Light-frame structures usually gain seismic resistance from rigid plywood shear walls and wood structural panel diaphragms. Special provisions for seismic load-resisting systems for all engineered wood structures requires consideration of diaphragm ratios, horizontal and vertical diaphragm shears, and connector/fastener values. In addition, collectors, or drag struts, to distribute shear along a diaphragm length are required.

Reinforced Masonry Structures

A construction system where steel reinforcement is embedded in the mortar joints of masonry or placed in holes and after filled with concrete or grout is called reinforced masonry.

The devastating 1933 Long Beach earthquake revealed that masonry construction should be improved immediately. Then, the California State Code made the reinforced masonry mandatory.

There are various practices and techniques to achieve reinforced masonry. The most common type is the reinforced hollow unit masonry. The effectiveness of both vertical and horizontal reinforcement strongly depends on the type and quality of the masonry, i.e. masonry units and mortar.

To achieve a ductile behavior of masonry, it is necessary that the shear strength of the wall is greater than the flexural strength.

Reinforced Concrete Structures

Stressed Ribbon pedestrian bridge over the Rogue River, Grants Pass, Oregon

Prestressed concrete cable-stayed bridge over Yangtze river

Reinforced concrete is concrete in which steel reinforcement bars (rebars) or fibers have been incorporated to strengthen a material that would otherwise be brittle. It can be used to produce beams, columns, floors or bridges.

Prestressed concrete is a kind of reinforced concrete used for overcoming concrete's natural weakness in tension. It can be applied to beams, floors or bridges with a longer span than is practical with ordinary reinforced concrete. Prestressing tendons (generally of high tensile steel cable or rods) are used to provide a clamping load which produces a compressive stress that offsets the tensile stress that the concrete compression member would, otherwise, experience due to a bending load.

To prevent catastrophic collapse in response earth shaking (in the interest of life safety), a traditional reinforced concrete frame should have ductile joints. Depending upon the methods used and the imposed seismic forces, such buildings may be immediately usable, require extensive repair, or may have to be demolished.

Prestressed Structures

Prestressed structure is the one whose overall integrity, stability and security depend, primarily, on a *prestressing*. *Prestressing* means the intentional creation of permanent stresses in a structure for the purpose of improving its performance under various service conditions.

Naturally pre-compressed exterior wall of Colosseum, Rome

There are the following basic types of prestressing:

- Pre-compression (mostly, with the own weight of a structure)

- Pretensioning with high-strength embedded tendons

- Post-tensioning with high-strength bonded or unbonded tendons

Today, the concept of prestressed structure is widely engaged in design of buildings, underground structures, TV towers, power stations, floating storage and offshore facilities, nuclear reactor vessels, and numerous kinds of bridge systems.

A beneficial idea of *prestressing* was, apparently, familiar to the ancient Rome architects; look, e.g., at the tall attic wall of Colosseum working as a stabilizing device for the wall piers beneath.

Steel Structures

Steel structures are considered mostly earthquake resistant but some failures have occurred. A great number of welded steel moment-resisting frame buildings, which looked earthquake-proof, surprisingly experienced brittle behavior and were hazardously damaged in the 1994 Northridge earthquake. After that, the Federal Emergency Management Agency (FEMA) initiated development of repair techniques and new design approaches to minimize damage to steel moment frame buildings in future earthquakes.

For structural steel seismic design based on Load and Resistance Factor Design (LRFD) approach, it is very important to assess ability of a structure to develop and maintain its bearing resistance in the inelastic range. A measure of this ability is ductility, which may be observed in a *material itself*, in a *structural element*, or to a *whole structure*.

As a consequence of Northridge earthquake experience, the American Institute of Steel Construction has introduced AISC 358 "Pre-Qualified Connections for Special and intermediate Steel Moment Frames." The AISC Seismic Design Provisions require that all Steel Moment Resisting Frames employ either connections contained in AISC 358, or the use of connections that have been subjected to pre-qualifying cyclic testing.

Prediction of Earthquake Losses

Earthquake loss estimation is usually defined as a *Damage Ratio* (DR) which is a ratio of the earthquake damage repair cost to the total value of a building. *Probable Maximum Loss* (PML) is a common term used for earthquake loss estimation, but it lacks a precise definition. In 1999, ASTM E2026 'Standard Guide for the Estimation of Building Damageability in Earthquakes' was produced in order to standardize the nomenclature for seismic loss estimation, as well as establish guidelines as to the review process and qualifications of the reviewer.

Earthquake loss estimations are also referred to as *Seismic Risk Assessments*. The risk assessment process generally involves determining the probability of various ground motions coupled with the vulnerability or damage of the building under those ground motions. The results are defined as a percent of building replacement value.

References

- Bozorgnia, Yousef; Bertero, Vitelmo V. (2004). Earthquake Engineering: From Engineering Seismology to Performance-Based Engineering. CRC Press. ISBN 978-0-8493-1439-1.

- Lindeburg, Michael R.; Baradar, Majid (2001). Seismic Design of Building Structures. Professional Publications. ISBN 1-888577-52-5.

- Chu, S.Y.; Soong, T.T.; Reinhorn, A.M. (2005). Active, Hybrid and Semi-Active Structural Control. John Wiley & Sons. ISBN 0-470-01352-4.

- Arnold, Christopher; Reitherman, Robert (1982). Building Configuration & Seismic Design. A Wiley-Interscience Publication. ISBN 0-471-86138-3.

- Reitherman, Robert (2012). Earthquakes and Engineers: An International History. Reston, VA: ASCE Press. pp. 394–395. ISBN 9780784410714.

- Craig Taylor; Erik VanMarcke, eds. (2002). Acceptable Risk Processes: Lifeline and Natural Hazards. Reston, VA: ASCE, TCLEE. ISBN 9780784406236.

- "Building Technology + Seismic Isolation System - OKUMURA CORPORATION" (in Japanese). Okumuragumi.co.jp. Retrieved 2012-07-31.

- "CMMI - Funding - Hazard Mitigation and Structural Engineering - US National Science Foundation (NSF)". nsf.gov. Retrieved 2012-07-31.

- "4. Building for earthquake resistance - Earthquakes - Te Ara Encyclopedia of New Zealand". Teara.govt.nz. 2009-03-02. Retrieved 2012-07-31.

Earthquake: An Overview

Earthquakes are a result of seismic activity in the Earth's crust. In this chapter the reader is informed about the causes of earthquakes and its various types such as megathrust earthquake, intraplate earthquake, interplate earthquake, deep-Focus earthquake and volcano tectonic earthquake. The features of each type of earthquake have been discussed to enable an insightful understanding of the subject of earthquakes.

Earthquake

An earthquake (also known as a quake, tremor or temblor) is the perceptible shaking of the surface of the Earth, resulting from the sudden release of energy in the Earth's crust that creates seismic waves. Earthquakes can be violent enough to toss people around and destroy whole cities. The seismicity or seismic activity of an area refers to the frequency, type and size of earthquakes experienced over a period of time.

Preliminary Determination of Epicenters
358,214 Events, 1963 - 1998

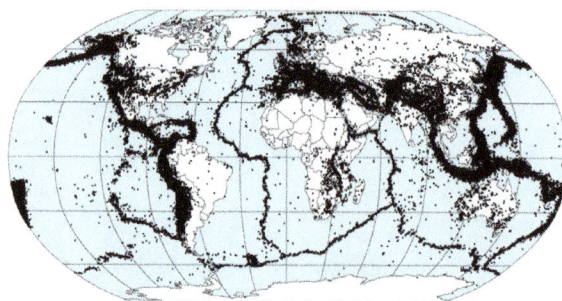

Global earthquake epicenters, 1963–1998

Earthquakes are measured using observations from seismometers. The moment magnitude is the most common scale on which earthquakes larger than approximately 5 are reported for the entire globe. The more numerous earthquakes smaller than magnitude 5 reported by national seismological observatories are measured mostly on the local magnitude scale, also referred to as the Richter magnitude scale. These two scales are numerically similar over their range of validity. Magnitude 3 or lower earthquakes are mostly imperceptible or weak and magnitude 7 and over potentially cause serious damage over larger areas, depending on their depth. The largest earthquakes in historic times have been of magnitude slightly over 9, although there is no limit to the possible magnitude. Intensity of shaking is measured on the modified Mercalli scale. The shallower an earthquake, the more damage to structures it causes, all else being equal.

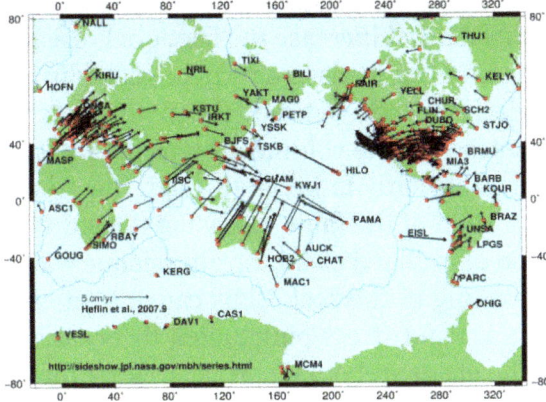

Global plate tectonic movement

At the Earth's surface, earthquakes manifest themselves by shaking and sometimes displacement of the ground. When the epicenter of a large earthquake is located offshore, the seabed may be displaced sufficiently to cause a tsunami. Earthquakes can also trigger landslides, and occasionally volcanic activity.

In its most general sense, the word *earthquake* is used to describe any seismic event — whether natural or caused by humans — that generates seismic waves. Earthquakes are caused mostly by rupture of geological faults, but also by other events such as volcanic activity, landslides, mine blasts, and nuclear tests. An earthquake's point of initial rupture is called its focus or hypocenter. The epicenter is the point at ground level directly above the hypocenter.

Naturally Occurring Earthquakes

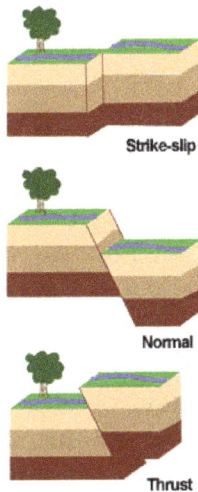

Fault types

Tectonic earthquakes occur anywhere in the earth where there is sufficient stored elastic strain energy to drive fracture propagation along a fault plane. The sides of a fault move past each other smoothly and aseismically only if there are no irregularities or asperities along the fault surface that increase the frictional resistance. Most fault surfaces do have such asperities and this leads to a form of stick-slip behavior. Once the fault has locked, continued relative motion between the plates leads to increasing stress and therefore, stored strain energy in the volume around the fault surface. This continues until the stress has risen sufficiently to break through the asperity, suddenly allowing sliding over the locked portion of the fault, releasing the stored energy. This energy is released as a combination of radiated elastic strain seismic waves, frictional heating of the fault surface, and cracking of the rock, thus causing an earthquake. This process of gradual build-up of strain and stress punctuated by occasional sudden earthquake failure is referred to as the elastic-rebound theory. It is estimated that only 10 percent or less of an earthquake's total energy is radiated as seismic energy. Most of the earthquake's energy is used to power the earthquake fracture growth or is converted into heat generated by friction. Therefore, earthquakes lower the Earth's available elastic potential energy and raise its temperature, though these changes are negligible compared to the conductive and convective flow of heat out from the Earth's deep interior.

Earthquake Fault Types

There are three main types of fault, all of which may cause an interplate earthquake: normal, reverse (thrust) and strike-slip. Normal and reverse faulting are examples of dip-slip, where the displacement along the fault is in the direction of dip and movement on them involves a vertical component. Normal faults occur mainly in areas where the crust is being extended such as a divergent boundary. Reverse faults occur in areas where the crust is being shortened such as at a convergent boundary. Strike-slip faults are steep structures where the two sides of the fault slip horizontally past each other; transform boundaries are a particular type of strike-slip fault. Many earthquakes are caused by movement on faults that have components of both dip-slip and strike-slip; this is known as oblique slip.

Reverse faults, particularly those along convergent plate boundaries are associated with the most powerful earthquakes, megathrust earthquakes, including almost all of those of magnitude 8 or more. Strike-slip faults, particularly continental transforms, can produce major earthquakes up to about magnitude 8. Earthquakes associated with normal faults are generally less than magnitude 7. For every unit increase in magnitude, there is a roughly thirtyfold increase in the energy released. For instance, an earthquake of magnitude 6.0 releases approximately 30 times more energy than a 5.0 magnitude earthquake and a 7.0 magnitude earthquake releases 900 times (30 × 30) more energy than a 5.0 magnitude of earthquake. An 8.6 magnitude earthquake releases the same amount of energy as 10,000 atomic bombs like those used in World War II.

This is so because the energy released in an earthquake, and thus its magnitude, is proportional to the area of the fault that ruptures and the stress drop. Therefore, the longer the length and the wider the width of the faulted area, the larger the resulting magnitude. The topmost, brittle part of the Earth's crust, and the cool slabs of the tectonic plates that are descending down into the hot mantle, are the only parts of our planet which can store elastic energy and release it in fault ruptures. Rocks hotter than about 300 degrees Celsius flow in response to stress; they do not rupture in earthquakes. The maximum observed lengths of ruptures and mapped faults (which may break in a single rupture) are approximately 1000 km. Examples are the earthquakes in Chile, 1960; Alaska, 1957; Sumatra, 2004, all in subduction zones. The longest earthquake ruptures on strike-slip faults, like the San Andreas Fault (1857, 1906), the North Anatolian Fault in Turkey (1939) and the Denali Fault in Alaska (2002), are about half to one third as long as the lengths along subducting plate margins, and those along normal faults are even shorter.

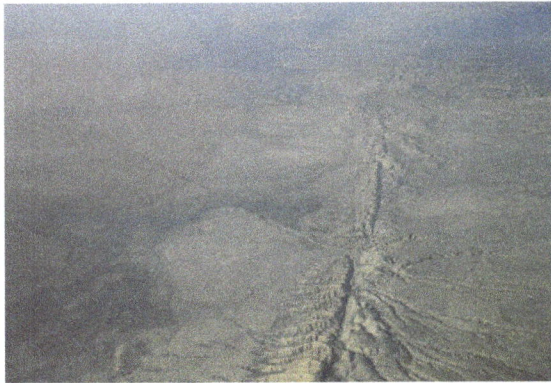

Aerial photo of the San Andreas Fault in the Carrizo Plain, northwest of Los Angeles

The most important parameter controlling the maximum earthquake magnitude on a fault is however not the maximum available length, but the available width because the latter varies by a factor of 20. Along converging plate margins, the dip angle of the rupture plane is very shallow, typically about 10 degrees. Thus the width of the plane within the top brittle crust of the Earth can become 50 to 100 km (Japan, 2011; Alaska, 1964), making the most powerful earthquakes possible.

Strike-slip faults tend to be oriented near vertically, resulting in an approximate width of 10 km within the brittle crust, thus earthquakes with magnitudes much larger than 8 are not possible. Maximum magnitudes along many normal faults are even more limited because many of them are located along spreading centers, as in Iceland, where the thickness of the brittle layer is only about 6 km.

In addition, there exists a hierarchy of stress level in the three fault types. Thrust faults are generated by the highest, strike slip by intermediate, and normal faults by the lowest stress levels. This can easily be understood by considering the direction of the greatest principal stress, the direction of the force that 'pushes' the rock mass during the faulting. In the case of normal faults, the rock mass is pushed down in a vertical direction, thus the pushing force (greatest principal stress) equals the weight of the rock mass itself. In the case of thrusting, the rock mass 'escapes' in the direction of the least principal stress, namely upward, lifting the rock mass up, thus the overburden equals the least principal stress. Strike-slip faulting is intermediate between the other two types described above. This difference in stress regime in the three faulting environments can contribute to differences in stress drop during faulting, which contributes to differences in the radiated energy, regardless of fault dimensions.

Earthquakes Away from Plate Boundaries

Where plate boundaries occur within the continental lithosphere, deformation is spread out over a much larger area than the plate boundary itself. In the case of the San Andreas fault continental transform, many earthquakes occur away from the plate boundary and are related to strains developed within the broader zone of deformation caused by major irregularities in the fault trace (e.g., the "Big bend" region). The Northridge earthquake was associated with movement on a blind thrust within such a zone. Another example is the strongly oblique convergent plate boundary between the Arabian and Eurasian plates where it runs through the northwestern part of the Zagros Mountains. The deformation associated with this plate boundary is partitioned into nearly pure thrust sense movements perpendicular to the boundary over a wide zone to the southwest and nearly pure strike-slip motion along the Main Recent Fault close to the actual plate boundary itself. This is demonstrated by earthquake focal mechanisms.

All tectonic plates have internal stress fields caused by their interactions with neighboring plates and sedimentary loading or unloading (e.g. deglaciation). These stresses may be sufficient to cause failure along existing fault planes, giving rise to intraplate earthquakes.

Shallow-focus and Deep-focus Earthquakes

Collapsed Gran Hotel building in the San Salvador metropolis,
after the shallow 1986 San Salvador earthquake.

The majority of tectonic earthquakes originate at the ring of fire in depths not exceeding tens of kilometers. Earthquakes occurring at a depth of less than 70 km are classified as 'shallow-focus' earthquakes, while those with a focal-depth between 70 and 300 km are commonly termed 'mid-focus' or 'intermediate-depth' earthquakes. In subduction zones, where older and colder oceanic crust descends beneath another tectonic plate, Deep-focus earthquakes may occur at much greater depths (ranging from 300 up to 700 kilometers). These seismically active areas of subduction are known as Wadati–Benioff zones. Deep-focus earthquakes occur at a depth where the subducted lithosphere should no longer be brittle, due to the high temperature and pressure. A possible mechanism for the generation of deep-focus earthquakes is faulting caused by olivine undergoing a phase transition into a spinel structure.

Earthquakes and Volcanic Activity

Earthquakes often occur in volcanic regions and are caused there, both by tectonic faults and the movement of magma in volcanoes. Such earthquakes can serve as an early warning of volcanic eruptions, as during the 1980 eruption of Mount St. Helens. Earthquake swarms can serve as markers for the location of the flowing magma throughout the volcanoes. These swarms can be recorded by seismometers and tiltmeters (a device that measures ground slope) and used as sensors to predict imminent or upcoming eruptions.

Rupture Dynamics

A tectonic earthquake begins by an initial rupture at a point on the fault surface, a process known as nucleation. The scale of the nucleation zone is uncertain, with some evidence, such as the rupture dimensions of the smallest earthquakes, suggesting that it is smaller than 100 m while other evidence, such as a slow component revealed by low-frequency spectra of some earthquakes, suggest that it is larger. The possibility that the nucleation involves some sort of preparation process is supported by the observation that about 40% of earthquakes are preceded by foreshocks. Once the rupture has initiated, it begins to propagate along the fault surface. The mechanics of this process are poorly understood, partly because it is difficult to recreate the high sliding velocities in a laboratory. Also the effects of strong ground motion make it very difficult to record information close to a nucleation zone.

Rupture propagation is generally modeled using a fracture mechanics approach, likening the rupture to a propagating mixed mode shear crack. The rupture velocity is a function of the fracture energy in the volume around the crack tip, increasing with decreasing fracture energy. The velocity of rupture propagation is orders of magnitude faster than the displacement velocity across the fault. Earthquake ruptures typically propagate at velocities that are in the range 70–90% of the S-wave velocity, and this is independent of earthquake size. A small subset of earthquake ruptures appear to have propagated at speeds greater than the S-wave velocity. These supershear earthquakes have all been observed during large strike-slip events. The unusually wide zone of coseismic damage caused by the 2001 Kunlun earthquake has been attributed to the effects of the sonic boom developed in such earthquakes. Some earthquake ruptures travel at unusually low velocities and are referred to as slow earthquakes. A particularly dangerous form of slow earthquake is the tsunami earthquake, observed where the relatively low felt intensities, caused by the slow propagation speed of some great earthquakes, fail to alert the population of the neighboring coast, as in the 1896 Sanriku earthquake.

Tidal Forces

Tides may induce some seismicity.

Earthquake Clusters

Most earthquakes form part of a sequence, related to each other in terms of location and time. Most earthquake clusters consist of small tremors that cause little to no damage, but there is a theory that earthquakes can recur in a regular pattern.

Aftershocks

An aftershock is an earthquake that occurs after a previous earthquake, the mainshock. An aftershock is in the same region of the main shock but always of a smaller magnitude. If an aftershock is larger than the main shock, the aftershock is redesignated as the main shock and the original main shock is redesignated as a foreshock. Aftershocks are formed as the crust around the displaced fault plane adjusts to the effects of the main shock.

Earthquake Swarms

Earthquake swarms are sequences of earthquakes striking in a specific area within a short period of time. They are different from earthquakes followed by a series of aftershocks by the fact that no single earthquake in the sequence is obviously the main shock, therefore none have notable higher magnitudes than the other. An example of an earthquake swarm is the 2004 activity at Yellowstone National Park. In August 2012, a swarm of earthquakes shook Southern California's Imperial Valley, showing the most recorded activity in the area since the 1970s.

Sometimes a series of earthquakes occur in what has been called an *earthquake storm*, where the earthquakes strike a fault in clusters, each triggered by the shaking or stress redistribution of the previous earthquakes. Similar to aftershocks but on adjacent segments of fault, these storms occur over the course of years, and with some of the later earthquakes as damaging as the early ones. Such a pattern was observed in the sequence of about a dozen earthquakes that struck the North Anatolian Fault in Turkey in the 20th century and has been inferred for older anomalous clusters of large earthquakes in the Middle East.

Size and Frequency of Occurrence

It is estimated that around 500,000 earthquakes occur each year, detectable with current instrumentation. About 100,000 of these can be felt. Minor earthquakes occur nearly constantly around the world in places like California and Alaska in the U.S., as well as in El Salvador, Mexico, Guatemala, Chile, Peru, Indonesia, Iran, Pakistan, the Azores in Portugal, Turkey, New Zealand, Greece, Italy, India, Nepal and Japan, but earthquakes can occur almost anywhere, including Downstate New York, England, and Australia. Larger earthquakes occur less frequently, the relationship being exponential; for example, roughly ten times as many earthquakes larger than magnitude 4 occur in a particular time period than earthquakes larger than magnitude 5. In the (low seismicity) United Kingdom, for example, it has been calculated that the average recurrences are: an earthquake of 3.7–4.6 every year, an earthquake of 4.7–5.5 every 10 years, and an earthquake of 5.6 or larger every 100 years. This is an example of the Gutenberg–Richter law.

The Messina earthquake and tsunami took as many as 200,000 lives on December 28, 1908 in Sicily and Calabria.

The number of seismic stations has increased from about 350 in 1931 to many thousands today. As a result, many more earthquakes are reported than in the past, but this is because of the vast improvement in instrumentation, rather than an increase in the number of earthquakes. The United States Geological Survey estimates that, since 1900, there have been an average of 18 major earthquakes (magnitude 7.0–7.9) and one great earthquake (magnitude 8.0 or greater) per year, and that this average has been relatively stable. In recent years, the number of major earthquakes per year has decreased, though this is probably a statistical fluctuation rather than a systematic trend. More detailed statistics on the size and frequency of earthquakes is available from the United States Geological Survey (USGS). A recent increase in the number of major earthquakes has been noted, which could be explained by a cyclical pattern of periods of intense tectonic activity, interspersed with longer periods of low-intensity. However, accurate recordings of earthquakes only began in the early 1900s, so it is too early to categorically state that this is the case.

Most of the world's earthquakes (90%, and 81% of the largest) take place in the 40,000 km long, horseshoe-shaped zone called the circum-Pacific seismic belt, known as the Pacific Ring of Fire, which for the most part bounds the Pacific Plate. Massive earthquakes tend to occur along other plate boundaries, too, such as along the Himalayan Mountains.

With the rapid growth of mega-cities such as Mexico City, Tokyo and Tehran, in areas of high seismic risk, some seismologists are warning that a single quake may claim the lives of up to 3 million people.

Induced Seismicity

While most earthquakes are caused by movement of the Earth's tectonic plates, human activity can also produce earthquakes. Four main activities contribute to this phenomenon: storing large amounts of water behind a dam (and possibly building an extremely heavy building), drilling and injecting liquid into wells, and by coal mining and oil drilling. Perhaps the best known example is the 2008 Sichuan earthquake in China's Sichuan Province in May; this tremor resulted in 69,227 fatalities and is the 19th deadliest earthquake of all time. The Zipingpu Dam is believed to have fluctuated the pressure of the fault 1,650 feet (503 m) away; this pressure probably increased the power of the earthquake and accelerated the rate of movement for the fault. The greatest earthquake in Australia's history is also claimed to be induced by humanity, through coal mining. The city of Newcastle was built over a large sector of coal mining areas. The earthquake has been reported to be spawned from a fault that reactivated due to the millions of tonnes of rock removed in the mining process.

Measuring and Locating Earthquakes

Earthquakes can be recorded by seismometers up to great distances, because seismic waves travel through the whole Earth's interior. The absolute magnitude of a quake is conventionally reported by numbers on the moment magnitude scale (formerly Richter scale, magnitude 7 causing serious damage over large areas), whereas the felt magnitude is reported using the modified Mercalli intensity scale (intensity II–XII). It has been proposed by researchers that, in order to measure the magnitude of an earthquake, one can use time difference between the arrival time of the peak high-frequency amplitude in an accelerogram (what records the acceleration of the ground during an earthquake) and the body-wave onset.

Every tremor produces different types of seismic waves, which travel through rock with different velocities:

- Longitudinal P-waves (shock- or pressure waves)
- Transverse S-waves (both body waves)
- Surface waves — (Rayleigh and Love waves)

Propagation velocity of the seismic waves ranges from approx. 3 km/s up to 13 km/s, depending on the density and elasticity of the medium. In the Earth's interior the shock- or P waves travel much faster than the S waves (approx. relation 1.7 : 1). The differences in travel time from the epicentre to the observatory are a measure of the distance and can be used to image both sources of quakes and structures within the Earth. Also the depth of the hypocenter can be computed roughly.

In solid rock P-waves travel at about 6 to 7 km per second; the velocity increases within the deep mantle to ~13 km/s. The velocity of S-waves ranges from 2–3 km/s in light sediments and 4–5 km/s in the Earth's crust up to 7 km/s in the deep mantle. As a consequence, the first waves of a distant earthquake arrive at an observatory via the Earth's mantle.

On average, the kilometer distance to the earthquake is the number of seconds between the P and S wave times 8. Slight deviations are caused by inhomogeneities of subsur- face structure. By such analyses of seismograms the Earth's core was located in 1913 by Beno Gutenberg.

Earthquakes are not only categorized by their magnitude but also by the place where they occur. The world is divided into 754 Flinn–Engdahl regions (F-E regions), which are based on political and geographical boundaries as well as seismic activity. More active zones are divided into smaller F-E regions whereas less active zones belong to larger F-E regions.

Standard reporting of earthquakes includes its magnitude, date and time of occurrence, geographic coordinates of its epicenter, depth of the epicenter, geographical region, distances to population centers, location uncertainty, a number of parameters that are included in USGS earthquake reports (number of stations reporting, number of obser- vations, etc.), and a unique event ID.

Effects of Earthquakes

1755 copper engraving depicting Lisbon in ruins and in flames after the 1755 Lisbon earthquake, which killed an estimated 60,000 people. A tsunami overwhelms the ships in the harbor.

The effects of earthquakes include, but are not limited to, the following:

Shaking and Ground Rupture

Damaged buildings in Port-au-Prince, Haiti, January 2010.

Shaking and ground rupture are the main effects created by earthquakes, principally resulting in more or less severe damage to buildings and other rigid structures. The severity of the local effects depends on the complex combination of the earthquake magnitude, the distance from the epicenter, and the local geological and geomorphological conditions, which may amplify or reduce wave propagation. The ground-shaking is measured by ground acceleration.

Specific local geological, geomorphological, and geostructural features can induce high levels of shaking on the ground surface even from low-intensity earthquakes. This effect is called site or local amplification. It is principally due to the transfer of the seismic motion from hard deep soils to soft superficial soils and to effects of seismic energy focalization owing to typical geometrical setting of the deposits.

Ground rupture is a visible breaking and displacement of the Earth's surface along the trace of the fault, which may be of the order of several metres in the case of major earthquakes. Ground rupture is a major risk for large engineering structures such as dams, bridges and nuclear power stations and requires careful mapping of existing faults to identify any which are likely to break the ground surface within the life of the structure.

Landslides and Avalanches

Earthquakes, along with severe storms, volcanic activity, coastal wave attack, and wildfires, can produce slope instability leading to landslides, a major geological hazard. Landslide danger may persist while emergency personnel are attempting rescue.

Fires

Earthquakes can cause fires by damaging electrical power or gas lines. In the event of water mains rupturing and a loss of pressure, it may also become difficult to stop the spread of a fire once it has started. For example, more deaths in the 1906 San Francisco earthquake were caused by fire than by the earthquake itself.

Fires of the 1906 San Francisco earthquake

Soil Liquefaction

Soil liquefaction occurs when, because of the shaking, water-saturated granular material (such as sand) temporarily loses its strength and transforms from a solid to a liquid. Soil liquefaction may cause rigid structures, like buildings and bridges, to tilt or sink into the liquefied deposits. For example, in the 1964 Alaska earthquake, soil liquefaction caused many buildings to sink into the ground, eventually collapsing upon themselves.

Tsunami

Tsunamis are long-wavelength, long-period sea waves produced by the sudden or abrupt movement of large volumes of water. In the open ocean the distance between wave crests can surpass 100 kilometers (62 mi), and the wave periods can vary from five minutes to one hour. Such tsunamis travel 600-800 kilometers per hour (373–497 miles per hour), depending on water depth. Large waves produced by an earthquake or a submarine landslide can overrun nearby coastal areas in a matter of minutes. Tsunamis can also travel thousands of kilometers across open ocean and wreak destruction on far shores hours after the earthquake that generated them.

Ordinarily, subduction earthquakes under magnitude 7.5 on the Richter scale do not cause tsunamis, although some instances of this have been recorded. Most destructive tsunamis are caused by earthquakes of magnitude 7.5 or more.

Floods

A flood is an overflow of any amount of water that reaches land. Floods occur usually when the volume of water within a body of water, such as a river or lake, exceeds the total capacity of the formation, and as a result some of the water flows or sits outside of the normal perimeter of the body. However, floods may be secondary effects of earthquakes, if dams are damaged. Earthquakes may cause landslips to dam rivers, which collapse and cause floods.

The terrain below the Sarez Lake in Tajikistan is in danger of catastrophic flood if the landslide dam formed by the earthquake, known as the Usoi Dam, were to fail during a future earthquake. Impact projections suggest the flood could affect roughly 5 million people.

Human Impacts

Ruins of the Għajn Ħadid Tower, which collapsed in an earthquake in 1856

An earthquake may cause injury and loss of life, road and bridge damage, general property damage, and collapse or destabilization (potentially leading to future collapse) of buildings. The aftermath may bring disease, lack of basic necessities, mental consequences such as panic attacks, depression to survivors, and higher insurance premiums.

Major Earthquakes

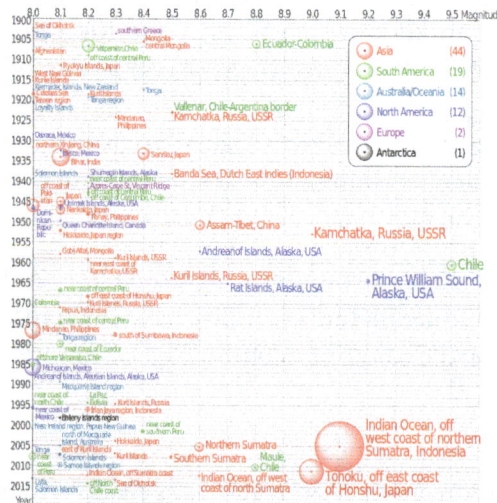

Earthquakes of magnitude 8.0 and greater since 1900. The apparent 3D volumes of the bubbles are linearly proportional to their respective fatalities.

One of the most devastating earthquakes in recorded history was the 1556 Shaanxi earthquake, which occurred on 23 January 1556 in Shaanxi province, China. More than 830,000 people died. Most houses in the area were yaodongs—dwellings carved out of loess hillsides—and many victims were killed when these structures collapsed. The 1976 Tangshan earthquake, which killed between 240,000 and 655,000 people, was the deadliest of the 20th century.

The 1960 Chilean earthquake is the largest earthquake that has been measured on a seismograph, reaching 9.5 magnitude on 22 May 1960. Its epicenter was near Cañete, Chile. The energy released was approximately twice that of the next most powerful earthquake, the Good Friday earthquake (March 27, 1964) which was centered in Prince William Sound, Alaska. The ten largest recorded earthquakes have all been megathrust earthquakes; however, of these ten, only the 2004 Indian Ocean earthquake is simultaneously one of the deadliest earthquakes in history.

Earthquakes that caused the greatest loss of life, while powerful, were deadly because of their proximity to either heavily populated areas or the ocean, where earthquakes often create tsunamis that can devastate communities thousands of kilometers away. Regions most at risk for great loss of life include those where earthquakes are relatively rare but powerful, and poor regions with lax, unenforced, or nonexistent seismic building codes.

Prediction

Many methods have been developed for predicting the time and place in which earthquakes will occur. Despite considerable research efforts by seismologists, scientifically reproducible predictions cannot yet be made to a specific day or month. However, for well-understood faults the probability that a segment may rupture during the next few decades can be estimated.

Earthquake warning systems have been developed that can provide regional notification of an earthquake in progress, but before the ground surface has begun to move, potentially allowing people within the system's range to seek shelter before the earthquake's impact is felt.

Preparedness

The objective of earthquake engineering is to foresee the impact of earthquakes on buildings and other structures and to design such structures to minimize the risk of damage. Existing structures can be modified by seismic retrofitting to improve their resistance to earthquakes. Earthquake insurance can provide building owners with financial protection against losses resulting from earthquakes.

Emergency management strategies can be employed by a government or organization to mitigate risks and prepare for consequences.

Historical Views

Tremblement de terre en Italie. 540 ans avant J.-C. — L. Papirius
Cursor consul (d'après Lycosthène).

An image from a 1557 book

From the lifetime of the Greek philosopher Anaxagoras in the 5th century BCE to the 14th century CE, earthquakes were usually attributed to "air (vapors) in the cavities of the Earth." Thales of Miletus, who lived from 625–547 (BCE) was the only documented person who believed that earthquakes were caused by tension between the earth and water. Other theories existed, including the Greek philosopher Anaxamines' (585–526 BCE) beliefs that short incline episodes of dryness and wetness caused seismic activity. The Greek philosopher Democritus (460–371 BCE) blamed water in general for earthquakes. Pliny the Elder called earthquakes "underground thunderstorms."

Recent Studies

In recent studies, geologists claim that global warming is one of the reasons for increased seismic activity. According to these studies melting glaciers and rising sea levels disturb the balance of pressure on Earth's tectonic plates thus causing increase in the frequency and intensity of earthquakes.

Earthquakes in Culture

Mythology and Religion

In Norse mythology, earthquakes were explained as the violent struggling of the god Loki. When Loki, god of mischief and strife, murdered Baldr, god of beauty and light, he was punished by being bound in a cave with a poisonous serpent placed above his head dripping venom. Loki's wife Sigyn stood by him with a bowl to catch the poison, but whenever she had to empty the bowl the poison dripped on Loki's face, forcing him to jerk his head away and thrash against his bonds, which caused the earth to tremble.

In Greek mythology, Poseidon was the cause and god of earthquakes. When he was in a bad mood, he struck the ground with a trident, causing earthquakes and other calamities. He also used earthquakes to punish and inflict fear upon people as revenge.

In Japanese mythology, Namazu (鯰) is a giant catfish who causes earthquakes. Namazu lives in the mud beneath the earth, and is guarded by the god Kashima who restrains the fish with a stone. When Kashima lets his guard fall, Namazu thrashes about, causing violent earthquakes.

In Popular Culture

In modern popular culture, the portrayal of earthquakes is shaped by the memory of great cities laid waste, such as Kobe in 1995 or San Francisco in 1906. Fictional earthquakes tend to strike suddenly and without warning. For this reason, stories about earthquakes generally begin with the disaster and focus on its immediate aftermath, as in *Short Walk to Daylight* (1972), *The Ragged Edge* (1968) or *Aftershock: Earthquake in New York* (1999). A notable example is Heinrich von Kleist's classic novella, *The Earthquake in Chile*, which describes the destruction of Santiago in 1647. Haruki Murakami's short fiction collection After the Quake depicts the consequences of the Kobe earthquake of 1995.

The most popular single earthquake in fiction is the hypothetical "Big One" expected of California's San Andreas Fault someday, as depicted in the novels *Richter 10* (1996), *Goodbye California* (1977), *2012* (2009) and *San Andreas* (2015) among other works. Jacob M. Appel's widely anthologized short story, *A Comparative Seismology*, features a con artist who convinces an elderly woman that an apocalyptic earthquake is imminent.

Contemporary depictions of earthquakes in film are variable in the manner in which they reflect human psychological reactions to the actual trauma that can be caused to directly afflicted families and their loved ones. Disaster mental health response research emphasizes the need to be aware of the different roles of loss of family and key community members, loss of home and familiar surroundings, loss of essential supplies and services to maintain survival. Particularly for children, the clear availability of caregiving adults who are able to protect, nourish, and clothe them in the aftermath of the earthquake, and to help them make sense of what has befallen them has been shown even more important to their emotional and physical health than the simple giving of provisions. As was observed after other disasters involving destruction and loss of life and their media depictions, such as those of the 2001 World Trade Center Attacks or Hurricane Katrina—and has been recently observed in the 2010 Haiti earthquake, it is also important not to pathologize the reactions to loss and displacement or disruption of governmental administration and services, but rather to validate these reactions, to support constructive problem-solving and reflection as to how one might improve the conditions of those affected.

Megathrust Earthquake

Megathrust earthquakes occur at subduction zones at destructive convergent plate boundaries, where one tectonic plate is forced underneath another. These interplate earthquakes are the planet's most powerful, with moment magnitudes (M_w) that can exceed 9.0. Since 1900, all earthquakes of magnitude 9.0 or greater have been megathrust earthquakes. No other type of known terrestrial source of tectonic activity has produced earthquakes of this scale.

Terminology

During the rupture, one side of the fault is pushed upwards relative to the other, and it is this type of movement that is known as thrust. They are a type of dip-slip fault. A thrust fault is a reverse fault with a dip of 45° or less. Oblique-slip faults have significant components of different slip styles. The term *megathrust* does not have a widely accepted rigorous definition, but is used to refer to an extremely large thrust fault, typically formed at the plate interface along a subduction zone such as the Sunda megathrust. It is mostly American terminology.

Areas

The major subduction zone is associated with the Pacific and Indian Oceans and is responsible for the volcanic activity associated with the Pacific Ring of Fire. Since these earthquakes deform the ocean floor, they often generate a significant series of tsunami waves. They are known to produce intense shaking for periods of time that can last for up to a few minutes.

In Japan, the Nankai megathrust under the Nankai Trough is responsible for Nankai megathrust earthquakes and associated tsunamis.

Examples

Examples of megathrust earthquakes are listed in the following table.

Event	Estimated Magnitude (M_w)	Tectonic Plates Involved	Other Details/Notes
365 Crete earthquake	8.0+	African Plate subducting beneath the Aegean Sea Plate	• The quake generated a large tsunami in the eastern Mediterranean Sea and caused significant vertical displacement in the island of Crete.
869 Sanriku earthquake	8.6–9.0	Pacific Plate subducting beneath the Okhotsk Plate	• Slip length: 200 km over (125 mi over) • Slip width: 85 km over (53 mi over)

Event	Estimated Magnitude (M_w)	Tectonic Plates Involved	Other Details/Notes
1575 Valdivia earthquake	8.5	Nazca Plate subducting beneath the South American Plate	
1700 Cascadia earthquake	8.7–9.2	Juan de Fuca Plate subducting beneath the North American Plate	• Slip length: 1000 km (625 mi) • Slip motion: 20 m (60 ft)
1707 Hōei earthquake	8.6–9.3	Philippine Sea Plate subducting beneath the Eurasian Plate	• Duration: approximately 10 minutes • Slip length: maybe 600 and 700 km (370 and 435 mi)
1730 Valparaíso earthquake	8.7-9.0	Nazca Plate subducting beneath the South American Plate	
1737 Kamchatka earthquake	8.3–9.0	Pacific Plate subducting beneath the Okhotsk Plate	• Duration: 15 minutes • Depth: 40 km • Slip length: maybe 700 km over (435 mi over)
1755 Lisbon earthquake	8.5–9.0	Hypothesized to be part of a young subduction zone but origin still debated. Related to the Azores–Gibraltar Transform Fault	• Destroyed Lisbon and was followed by a 20 metre high tsunami and many fires. • It caused 10,000-100,000 deaths. • The grave of Nuno Álvares Pereira was destroyed, along with the historical records of the voyages of Vasco da Gama and Christopher Columbus
1868 Arica earthquake	8.5–9.0	Nazca Plate subducting beneath the South American Plate	• Slip length: 600 km (370 mi)
1877 Iquique earthquake	8.5–9.0?	Nazca Plate subducting beneath the South American Plate	• Slip length: 420 and 450 km (230 and 245 mi)
1906 Ecuador–Colombia earthquake	8.8	Nazca Plate subducting beneath the South American Plate	
1946 Nankaidō earthquake	8.1	Philippine Sea Plate subducting beneath the Eurasian Plate	• Slip length: maybe 300 km (190 mi)
1952 Kamchatka earthquake	9.0	Pacific Plate subducting beneath the Okhotsk Plate	• Depth: 30 km • Slip length: maybe 600 km over (370 mi over)
1957 Andreanof Islands earthquake	8.6	Pacific Plate subducting beneath the North American Plate	• Depth: 33 km • Slip length: maybe 700 km over (435 mi over)

Event	Estimated Magnitude (M_w)	Tectonic Plates Involved	Other Details/Notes
1960 Great Chilean earthquake	9.5	Nazca Plate subducting beneath the South American Plate	• Duration: 5–6 minutes • Depth: 33 km • Slip length: 850 and 1000 km (530 and 625 mi) • Slip width: 200 km (125 mi) • Slip motion: 20 m (60 ft)
1964 Alaska earthquake ("Good Friday" earthquake)	9.2	Pacific Plate subducting beneath the North American Plate	• Duration: 4–5 minutes • Depth: 25 km • Slip length: 800 and 850 km (500 and 530 mi) • Slip width: 250 km (155 mi) • Slip motion: 23 m (70 ft)
2001 southern Peru earthquake	8.4	Nazca Plate subducting beneath the South American Plate	• Depth: 33 km • Slip length: 200 km (125 mi) • Slip width: 110 km (70 mi)
2004 Sumatra-Andaman earthquake ("Indian Ocean earthquake")	9.1–9.3	India Plate subducting beneath the Burma Plate	• The total vertical displacement measured by sonar survey is about 40 m in the vicinity of the epicenter and occurred as two separate movements which created two large, steep, almost vertical cliffs, one above the other. • Duration: 8–10 minutes • Depth: 30 km • Slip length: 1000 and 1300 km (625 and 810 mi) • Slip width: 180 km (110 mi) • Slip motion: 33 m (110 ft)
2010 Chile earthquake	8.8	Nazca Plate subducting beneath the South American Plate	• Depth: 35 km • Slip length: 500 km (310 mi) • Slip width: 200 km (125 mi)
2011 Tōhoku earthquake and tsunami	9.0	Pacific Plate subducting beneath the Okhotsk Plate	• Duration: 6 minutes • Depth: 30 km • Slip length: 500 km (310 mi) • Slip width: 200 km (125 mi) • Slip motion: 20 m (60 ft)

Event	Estimated Magnitude (M_w)	Tectonic Plates Involved	Other Details/Notes
2014 Iquique earthquake	8.2	Nazca Plate subducting beneath the South American Plate	• Depth: 20.1 km • Slip length: 170 km (105 mi) • Slip width: 70 km (45 mi)
2015 Illapel earthquake	8.3	Nazca Plate subducting beneath the South American Plate	• Depth: 25 km • Slip length: 260 km (160 mi) • Slip width: 80 km (50 mi)

Intraplate Earthquake

An intraplate earthquake occurs in the interior of a tectonic plate, whereas an interplate earthquake is one that occurs at a plate boundary.

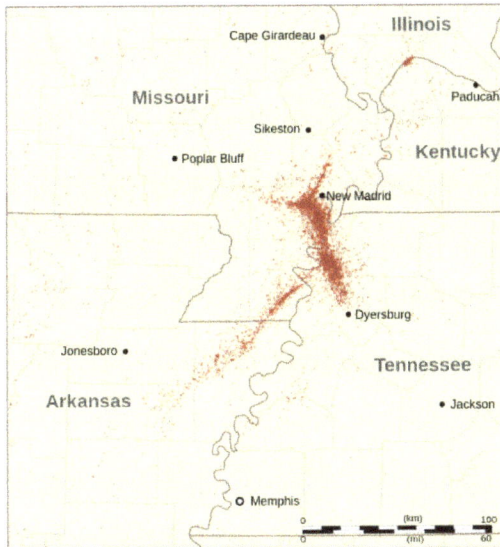

Distribution of seismicity associated with the New Madrid Seismic Zone (since 1974). This zone of intense earthquake activity is located deep in the interior of the North American plate.

Intraplate earthquakes are relatively rare. Interplate earthquakes, which occur at plate boundaries, are more common. Nonetheless, very large intraplate earthquakes can inflict heavy damage, particularly because such areas are not accustomed to earthquakes and buildings are usually not seismically retrofitted. Examples of damaging intraplate earthquakes are the devastating Gujarat earthquake in 2001, the 2012 Indian Ocean earthquakes, the 1811-1812 earthquakes in New Madrid, Missouri, and the 1886 earthquake in Charleston, South Carolina.

Fault Zones within Tectonic Plates

The surface of the Earth is made up of seven primary and eight secondary tectonic plates, plus dozens of tertiary microplates. The large plates move very slowly, owing to convection currents within the mantle below the crust. Because they do not all move in the same direction, plates often directly collide or move laterally along each other, a tectonic environment that makes earthquakes frequent. Relatively few earthquakes occur in intraplate environments; most occur on faults near plate margins. By definition, intraplate earthquakes do not occur near plate boundaries, but along faults in the normally stable interior of plates. These earthquakes often occur at the location of ancient failed rifts, because such old structures may present a weakness in the crust where it can easily slip to accommodate regional tectonic strain.

Compared to earthquakes near plate boundaries, intraplate earthquakes are not well understood, and the hazards associated with them may be difficult to quantify.

Historic Examples

Historic examples of intraplate earthquakes include those in Mineral, Virginia in 2011 (estimated magnitude 5.8), New Madrid in 1811 and 1812 (estimated magnitude as high as 8.1), the Boston (Cape Ann) earthquake of 1755 (estimated magnitude 6.0 to 6.3), earthquakes felt in New York City in 1737 and 1884 (both quakes estimated at about 5.5 magnitude), and the Charleston earthquake in South Carolina in 1886 (estimated magnitude 6.5 to 7.3). The Charleston quake was particularly surprising because, unlike Boston and New York, the area had almost no history of even minor earthquakes.

In 2001, a large intraplate earthquake devastated the region of Gujarat, India. The earthquake occurred far from any plate boundaries, which meant the region above the epicenter was unprepared for earthquakes. In particular, the Kutch district suffered tremendous damage, where the death toll was over 12,000 and the total death toll was higher than 20,000.

Causes

Many cities live with the seismic risk of a rare, large intraplate earthquake. The cause of these earthquakes is often uncertain. In many cases, the causative fault is deeply buried, and sometimes cannot even be found. Under these circumstances it is difficult to calculate the exact seismic hazard for a given city, especially if there was only one earthquake in historical times. Some progress is being made in understanding the fault mechanics driving these earthquakes.

Prediction

Scientists continue to search for the causes of these earthquakes, and especially for some indication of how often they recur. The best success has come with detailed micro-seismic monitoring, involving dense arrays of seismometers. In this manner, very small earthquakes associated with a causative fault can be located with great accuracy, and in most cases these line up in patterns consistent with faulting. Cryoseisms can sometimes be mistaken for intraplate earthquakes.

Interplate Earthquake

An interplate earthquake is an earthquake that occurs at the boundary between two tectonic plates. Earthquakes of this type account for more than 90 percent of the total seismic energy released around the world. If one plate is trying to move past the other, they will be locked until sufficient stress builds up to cause the plates to slip relative to each other. The slipping process creates an earthquake with land deformations and resulting seismic waves which travel through the Earth and along the Earth's surface. Relative plate motion can be lateral as along a transform fault boundary or vertical if along a convergent subduction boundary or a rift at a divergent boundary. At a subduction boundary the motion is due to one plate slipping beneath the other plate resulting in an interplate thrust or megathrust earthquake, which are the most powerful earthquakes.

Some areas of the world that are particularly prone to such events include the west coast of North America (especially California and Alaska), the northeastern Mediterranean region (Greece, Italy, and Turkey in particular), Iran, New Zealand, Indonesia, India, Japan, and parts of China.

Interplate earthquakes differ from intraplate earthquake in the intensity of stress drop which occurs after the quake. Intraplate earthquake have, on average, more stress drop than that of the interplate earthquake. Interplate earthquakes also differ fundamentally from intraplate earthquakes in the way stress is released and recovered. An interplate earthquake results in an immediate stress drop along the fault. Following this is a period of postseismic stress restoration. This restoration occurs quickly within the first few decades following the rupture and is due to tectonic loading and viscous relaxation in the lower crust. This results in a transfer of stress to the upper crust. Later on, a period of steady stress increase occurs due to tectonic loading.

Deep-focus Earthquake

A deep-focus earthquake in seismology is an earthquake with a hypocenter depth exceeding 300 km. They occur almost exclusively at oceanic-continental convergent boundaries in association with subducted oceanic lithosphere. They occur along a dipping tabular zone beneath the subduction zone known as the Wadati–Benioff zone.

Discovery

Preliminary evidence for the existence of deep-focus earthquakes was first brought to the attention of the scientific community in 1922 by Herbert Hall Turner. In 1928, Wadati proved the existence of earthquakes occurring well beneath the lithosphere, dispelling the notion that earthquakes occur only with shallow focal depths.

Seismic Characteristics

Deep-focus earthquakes give rise to minimal surface waves. Their focal depth causes the earthquakes to be less likely to produce seismic wave motion with energy concentrated at the surface. The path of deep-focus earthquake seismic waves from focus to recording station goes through the heterogeneous upper mantle and highly variable crust only once. Therefore, the body waves undergo less attenuation and reverberation than seismic waves from shallow earthquakes, resulting in sharp body wave peaks.

Focal Mechanisms

The pattern of energy radiation of an earthquake is represented by the moment tensor solution, which is graphically represented by beachball diagrams. An explosive or implosive mechanism produces an isotropic seismic source . Slip on a planar fault surface results in what is known as a double-couple source. Uniform outward motion in a single plane due to normal shortening gives rise is known as a compensated linear vector dipole source. Deep-focus earthquakes have been shown to contain a combination of these sources.

Physical Process

Shallow-focus earthquakes are the result of the sudden release of strain energy built up over time in rock by brittle fracture and frictional slip over planar surfaces. However, the physical mechanism of deep focus earthquakes is poorly understood. Subducted lithosphere subject to the pressure and temperature regime at depths greater than 300 km should not exhibit brittle behavior, but should rather respond to stress by plastic deformation. Several physical mechanisms have been proposed for the nucleation and propagation of deep-focus earthquakes; however, the exact process remains an outstanding problem in the field of deep earth seismology.

The following four subsections outline proposals which could explain the physical mechanism allowing deep focus earthquakes to occur. With the exception of solid-solid phase transitions, the proposed theories for the focal mechanism of deep earthquakes hold equal footing in current scientific literature.

Solid-solid Phase Transitions

The earliest proposed mechanism for the generation of deep-focus earthquakes is an implosion due to a phase transition of material to a higher density, lower volume phase. The olivine-spinel phase transition is thought to occur at a depth of 410 km in the interior of the earth. This hypothesis proposes that metastable olivine in oceanic lithosphere subducted to depths greater than 410 km undergoes a sudden phase transition to spinel structure. The increase in density due to the reaction would cause an implosion giving rise to the earthquake. This mechanism has been largely discredited due to the lack of a significant isotropic signature in the moment tensor solution of deep-focus earthquakes.

Dehydration Embrittlement

Dehydration reactions of mineral phases with high weight percent water would increase the pore pressure in a subducted oceanic lithosphere slab. This effect reduces the effective normal stress in the slab and allow slip to occur on pre-existing fault planes at significantly greater depths that would normally be possible. Several workers[who?] suggest that this mechanism does not play a significant role in seismic activity beyond 350 km depth due to the fact that most dehydration reactions will have reached completion by a pressure corresponding to 150 to 300 km depth (5-10 GPa).

Transformational Faulting or Anticrack Faulting

Transformational faulting, also known as anticrack faulting, is the result of the phase transition of a mineral to a higher density phase occurring in response to shear stress in a fine-grained shear zone. The transformation occurs along the plane of maximal shear stress. Rapid shearing can then occur along these planes of weakness, giving rise to an earthquake in a mechanism similar to a shallow-focus earthquake. Metastable olivine subducted past the olivine-wadsleyite transition at 320--410 km depth (depending on temperature) is a potential candidate for such instabilities. Arguments against this hypothesis include the requirements that the faulting region should be very cold, and contain very little mineral-bound hydroxyl. Higher temperatures or higher hydroxyl contents preclude the metastable preservation of olivine to the depths of the deepest earthquakes.

Shear Instability / Thermal Runaway

A shear instability arises when heat is produced by plastic deformation faster than it can be conducted away. The result is thermal runaway, a positive-feedback loop of heating, material weakening and strain-localisation within the shear zone. Continued weakening may result in partial melting along zones of maximal shear stress. Plastic shear instabilities leading to earthquakes have not been documented in nature, nor have they been observed in natural materials in the laboratory. Their relevance to deep earthquakes therefore lies in mathematical models which use simplified material properties and rheologies to simulate natural conditions.

Notable Deep-focus Earthquakes

The strongest deep-focus earthquake in seismic record was the 2013 Okhotsk Sea earthquake (magnitude 8.3) that occurred with an epicenter in the Sea of Okhotsk at a depth of 609 km. The deepest ever recorded earthquake is the 1994 Bolivia earthquake with a focal depth of 647 km and a moment magnitude of 8.2.

Volcano Tectonic Earthquake

A volcano tectonic earthquake is an earthquake induced by the movement (injection or withdrawal) of magma. The movement results in pressure changes in the rock around where the magma has experienced stress. At some point, the rock may break or move. The earthquakes may also be related to dike intrusion and may occur as earthquake swarms. An example is the 2007–2008 Nazko earthquake swarm in central British Columbia, Canada.

Other types of seismic activity related to volcanoes and their eruptions are long period seismic waves, which are from sudden sporadic movement of magma, which is blocked from moving due to a blockage. Another is a harmonic tremor, which is steady movement of magma, deep in the mantle.

Example of Volcanic Tectonic Earthquake

Subduction

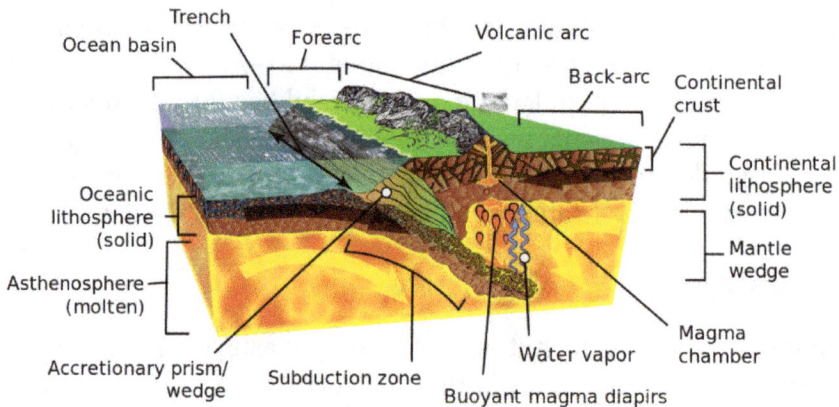

Diagram of the geological process of subduction

Subduction is a geological process that takes place at convergent boundaries of tectonic plates where one plate moves under another and is forced down into the mantle. Regions where this process occurs are known as *subduction zones*. Rates of subduction are typically in centimetres per year, with the average rate of convergence being approximately two to eight centimetres per year along most plate boundaries.

Plates include both oceanic crust and continental crust. Stable subduction zones involve the oceanic lithosphere of one plate sliding beneath the continental or oceanic lithosphere of another plate due to the higher density of the oceanic lithosphere. That is, the subducted lithosphere is always oceanic while the over-riding lithosphere may or may not be oceanic. Subduction zones are sites that have a high rate of volcanism, earthquakes, and mountain building.

Orogenesis, or mountain-building, occurs when large pieces of material on the subducting plate (such as island arcs) are pressed into the over-riding plate or when subhorizontal contraction occurs in the over-riding plate. These areas are subject to many earthquakes, which are caused by the interactions between the subducting slab and the mantle, the volcanoes, and (when applicable) the mountain-building related to island arc collisions.

General Description

Subduction zones are sites of convective downwelling of Earth's lithosphere (the crust plus the top non-convecting portion of the upper mantle). Subduction zones exist at convergent plate boundaries where one plate of oceanic lithosphere converges with another plate. The descending slab, the subducting plate, is over-ridden by the leading edge of the other plate. The slab sinks at an angle of approximately twenty-five to forty-five degrees to Earth's surface. This sinking is driven by the temperature difference between the subducting oceanic lithosphere and the surrounding mantle asthenosphere, as the colder oceanic lithosphere is, on average, more dense. At a depth of approximately 80–120 kilometres, the basalt of the oceanic crust is converted to a metamorphic rock called eclogite. At that point, the density of the oceanic crust increases and provides additional negative buoyancy (downwards force). It is at subduction zones that Earth's lithosphere, oceanic crust, sedimentary layers, and some trapped water are recycled into the deep mantle.

Earth is so far the only planet where subduction is known to occur. Subduction is the driving force behind plate tectonics, and without it, plate tectonics could not occur.

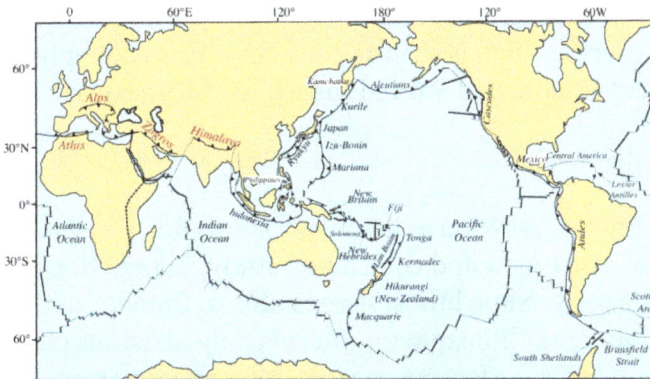

Global convergent plate margins, ~55,000 km long, shown as barbed lines. These are connected to subduction ones at depth. Convergent margins include continental *collision zones* as well as normal subduction zones.

Subduction zones dive down into the mantle beneath 55,000 kilometres of convergent plate margins (Lallemand, 1999), almost equal to the cumulative 60,000 kilometres of mid-ocean ridges. Subduction zones burrow deeply but are imperfectly camouflaged, and geophysics and geochemistry can be used to study them. Not surprisingly, the shallowest portions of subduction zones are known best. Subduction zones are strongly asymmetric for the first several hundred kilometres of their descent. They start to go down at oceanic trenches. Their descents are marked by inclined zones of earthquakes that dip away from the trench beneath the volcanoes and extend down to the 660-kilometre discontinuity. Subduction zones are defined by the inclined array of earthquakes known as the Wadati-Benioff zone after the two scientists who first identified this distinctive aspect. Subduction zone earthquakes occur at greater depths (up to 600 km) than elsewhere on Earth (typically <20 km depth); such deep earthquakes may be driven by deep phase transformations, thermal runaway, or dehydration embrittlement.

The subducting basalt and sediment are normally rich in hydrous minerals and clays. Additionally, large quantities of water are introduced into cracks and fractures created as the subducting slab bends downward. During the transition from basalt to eclogite, these hydrous materials break down, producing copious quantities of water, which at such great pressure and temperature exists as a supercritical fluid. The supercritical water, which is hot and more buoyant than the surrounding rock, rises into the overlying mantle where it lowers the pressure in (and thus the melting temperature of) the mantle rock to the point of actual melting, generating magma. The magmas, in turn, rise because they are less dense than the rocks of the mantle. The mantle-derived magmas (which are basaltic in composition) can continue to rise, ultimately to Earth's surface, resulting in a volcanic eruption. The chemical composition of the erupting lava depends upon the degree to which the mantle-derived basalt interacts with (melts) Earth's crust and/or undergoes fractional crystallization.

Above subduction zones, volcanoes exist in long chains called volcanic arcs. Volcanoes that exist along arcs tend to produce dangerous eruptions because they are rich in water (from the slab and sediments) and tend to be extremely explosive. Krakatoa, Nevado del Ruiz, and Mount Vesuvius are all examples of arc volcanoes. Arcs are also known to be associated with precious metals such as gold, silver and copper believed to be carried by water and concentrated in and around their host volcanoes in rock called "ore".

Theory on Origin

Although the process of subduction as it occurs today is fairly well understood, its origin remains a matter of discussion and continuing study. Subduction initiation can occur *spontaneously* if denser oceanic lithosphere is able to founder and sink beneath adjacent oceanic or continental lithosphere; alternatively, existing plate motions can *induce* new subduction zones by forcing oceanic lithosphere to rupture and sink into the asthenosphere. Both models can eventually yield self-sustaining subduction zones, as oceanic crust is metamorphosed at great depth and becomes denser than the surrounding mantle rocks. Results from numerical models generally favor induced subduction initiation for most modern subduction zones, but other analogue modeling shows the possibility of spontaneous subduction from inherent density differences between two plates at passive margins, and observations from the Izu-Bonin-Mariana subduction system are compatible with spontaneous subduction nucleation. Furthermore, subduction is likely to have spontaneously initiated at some point in Earth's history, as induced subduction nucleation requires existing plate motions, though an unorthodox proposal by A. Yin suggests that meteorite impacts may have contributed to subduction initiation on early Earth.

Geophysicist Don L. Anderson has hypothesized that plate tectonics could not happen without the calcium carbonate laid down by bioforms at the edges of subduction zones. The massive weight of these sediments could be softening the underlying rocks, making them pliable enough to plunge. However, considering that some refractory minerals used for dating early Earth, such as zircon, are typically generated in subduction zones and associated with granites and pegmatites, some of these early dates may have preceded significant biological activity on Earth.

Effects

Metamorphism

Volcanic Activity

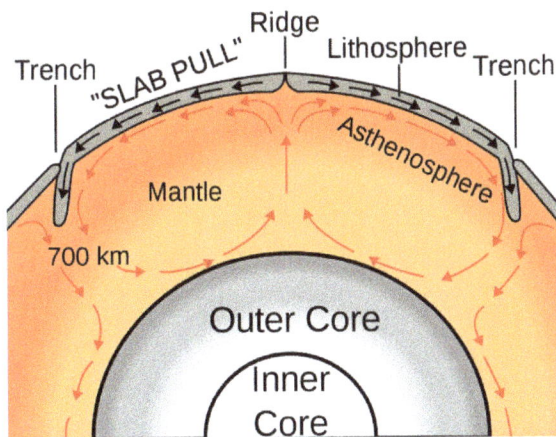

Oceanic plates are subducted creating oceanic trenches.

Volcanoes that occur above subduction zones, such as Mount St. Helens, Mount Etna and Mount Fuji, lie at approximately one hundred kilometres from the trench in arcuate chains, hence the term volcanic arc. Two kinds of arcs are generally observed on Earth: island arcs that form on oceanic lithosphere (for example, the Mariana and the Tonga island arcs), and continental arcs such as the Cascade Volcanic Arc, that form along the coast of continents. Island arcs are produced by the subduction of oceanic lithosphere beneath another oceanic lithosphere (oceanic subduction) while continental arcs formed during subduction of oceanic lithosphere beneath a continental lithosphere.

The arc magmatism occurs one hundred to two hundred kilometres from the trench and approximately one hundred kilometres from the subducting slab. This depth of arc magma generation is the consequence of the interaction between fluids, released from the subducting slab, and the arc mantle wedge that is hot enough to generate hydrous melting. Arcs produce about 25% of the total volume of magma produced each year on Earth (approximately thirty to thirty-five cubic kilometres), much less than the volume produced at mid-ocean ridges, and they contribute to the formation of new continental crust. Arc volcanism has the greatest impact on humans, because many arc volcanoes lie above sea level and erupt violently. Aerosols injected into the stratosphere during violent eruptions can cause rapid cooling of Earth's climate and affect air travel.

Earthquakes and Tsunamis

The strains caused by plate convergence in subduction zones cause at least three different types of earthquakes. Earthquakes mainly propagate in the cold subducting slab and define the Wadati-Benioff zone. Seismicity shows that the slab can be tracked down to the upper mantle/lower mantle boundary (approximately six hundred kilometre depth).

Nine out of the ten largest earthquakes to occur in the last century were subduction zone events, which includes the 1960 Great Chilean earthquake, which, at M 9.5, was the largest earthquake ever recorded; the 2004 Indian Ocean earthquake and tsunami; and the 2011 Tōhoku earthquake and tsunami. The subduction of cold oceanic crust into the mantle depresses the local geothermal gradient and causes a larger portion of Earth to deform in a more brittle fashion than it would in a normal geothermal gradient setting. Because earthquakes can occur only when a rock is deforming in a brittle fashion, subduction zones can cause large earthquakes. If such a quake causes rapid deformation of the sea floor, there is potential for tsunamis, such as the earthquake caused by subduction of the Indo-Australian Plate under the Euro-Asian Plate on December 26, 2004 that devastated the areas around the Indian Ocean. Small tremors, with cause small, nondamaging tsunamis, occur frequently.

Outer rise earthquakes occur when normal faults oceanward of the subduction zone are activated by flexture of the plate as it bends into the subduction zone. The Samoa earthquake of 2009 is an example of this type of event. Displacement of the sea floor caused by this event generated a six-metre tsunami in nearby Samoa.

Anomalously deep events are a characteristic of subduction zones, which produce the deepest quakes on the planet. Earthquakes are generally restricted to the shallow, brittle parts of the crust, generally at depths of less than twenty kilometres. However, in subduction zones, quakes occur at depths as great as seven hundred kilometres. These quakes define inclined zones of seismicity known as Wadati-Benioff zones, after the scientists who discovered them, which trace the descending lithosphere. Seismic tomography has helped detect subducted lithosphere in regions where there are no earthquakes. Some subducted slabs seem not to be able to penetrate the major discontinuity in the mantle that lies at a depth of about 670 kilometres whereas other subducted oceanic plates can penetrate all the way to the core-mantle boundary. The great seismic discontinuities in the mantle, at 410 and 670 kilometre depth, are disrupted by the descent of cold slabs in deep subduction zones.

Orogeny

Subducting plates can bring island arcs and sediments to convergent margins. The material often does not subduct with the rest of the plate but instead is accreted to the continent in the form of exotic terranes. They cause crustal thickening and mountain-building. This accretion process is thought by many geologists to be the source of much of western North America and of the uplift that produced the Rocky Mountains.

Subduction Angle

Subduction typically occurs at a moderately steep angle right at the point of the convergent plate boundary. However, anomalous shallower angles of subduction are known to exist as well some that are extremely steep.

- Flat-slab subduction (<30°): occurs when subducting lithosphere, called a slab, subducts horizontally or nearly horizontally. The flat slab can extend for hundreds of kilometres and can even extend to over a thousand. That is abnormal, as the dense slab typically sinks at a much steeper angle directly at the subduction zone. Because subduction of slabs to depth is necessary to drive subduction zone volcanism (through the destabilization and dewatering of minerals and the resultant flux melting of the mantle wedge), flat-slab subduction can be invoked to explain volcanic gaps. Flat-slab subduction is ongoing beneath part of the Andes causing segmentation of the Andean Volcanic Belt into four zones. The flat-slab subduction in northern Peru and Norte Chico region of Chile is believed to be the result of the subduction of two buoyant aseismic ridges, the Nazca Ridge and the Juan Fernández Ridge respectively. Around Taitao Peninsula flat-slab subduction is attributed to the subduction of the Chile Rise, a spreading ridge. The Laramide Orogeny in the Rocky Mountains of United States is attributed to flat-slab subduction. Then, a broad volcanic gap appeared at the southwestern margin of North America, and deformation occurred much farther inland; it was during this time that the basement-cored mountain ranges of Colorado, Utah, Wyoming, South Dakota, and New Mexico came into being.

- Steep-angle subduction (>70°): occurs in subduction zones where Earth's oceanic crust and lithosphere are old and thick and have, therefore, lost buoyancy.

The steepest dipping subduction zone lies in the Mariana Trench, which is also where the oceanic crust, of Jurassic age, is the oldest on Earth exempting ophiolites. Steep-angle subduction is, in contrast to flat-slab subduction, associated with back-arc extension of crust making volcanic arcs and fragments of continental crust wander away from continents over geological times leaving behind a marginal sea.

Importance

Subduction zones are important for several reasons:

1. Subduction Zone Physics: Sinking of the oceanic lithosphere (sediments, crust, mantle), by contrast of density between the cold and old lithosphere and the hot asthenospheric mantle wedge, is the strongest force (but not the only one) needed to drive plate motion and is the dominant mode of mantle convection.

2. Subduction Zone Chemistry: The subducted sediments and crust dehydrate and release water-rich (aqueous) fluids into the overlying mantle, causing mantle melting and fractionation of elements between surface and deep mantle reservoirs, producing island arcs and continental crust.

3. Subduction zones drag down subducted oceanic sediments, oceanic crust, and mantle lithosphere that interact with the hot asthenospheric mantle from the over-riding plate to produce calc-alkaline series melts, ore deposits, and continental crust.

Subduction zones have also been considered as possible disposal sites for nuclear waste in which the action of subduction itself would carry the material into the planetary mantle, safely away from any possible influence on humanity or the surface environment. However, that method of disposal is currently banned by international agreement. Furthermore, plate subduction zones are associated with very large megathrust earthquakes, making the effects on using any specific site for disposal unpredictable and possibly adverse to the safety of longterm disposal.

References

- Bozorgnia, Yousef; Bertero, Vitelmo V. (2004). Earthquake Engineering: From Engineering Seismology to Performance-Based Engineering. CRC Press. ISBN 978-0-8493-1439-1.

- Lindeburg, Michael R.; Baradar, Majid (2001). Seismic Design of Building Structures. Professional Publications. ISBN 1-888577-52-5.

- Chu, S.Y.; Soong, T.T.; Reinhorn, A.M. (2005). Active, Hybrid and Semi-Active Structural Control. John Wiley & Sons. ISBN 0-470-01352-4.

- Arnold, Christopher; Reitherman, Robert (1982). Building Configuration & Seismic Design. A Wiley-Interscience Publication. ISBN 0-471-86138-3.

- Robert W. Day (2007). Geotechnical Earthquake Engineering Handbook. McGraw Hill. ISBN 0-07-137782-4.

- Reitherman, Robert (2012). Earthquakes and Engineers: An International History. Reston, VA: ASCE Press. pp. 394–395. ISBN 9780784410714.

- EERI Endowment Subcommittee (May 2000). Financial Management of Earthquake Risk. EERI Publication. ISBN 0-943198-21-6.

- Craig Taylor; Erik VanMarcke, eds. (2002). Acceptable Risk Processes: Lifeline and Natural Hazards. Reston, VA: ASCE, TCLEE. ISBN 9780784406236.

- "Building Technology + Seismic Isolation System - OKUMURA CORPORATION" (in Japanese). Okumuragumi.co.jp. Retrieved 2012-07-31.

- "Strategy to Close Metsamor Plant Presented | Asbarez Armenian News". Asbarez.com. 1995-10-26. Retrieved 2012-07-31.

- "CMMI - Funding - Hazard Mitigation and Structural Engineering - US National Science Foundation (NSF)". nsf.gov. Retrieved 2012-07-31.

- "4. Building for earthquake resistance - Earthquakes - Te Ara Encyclopedia of New Zealand". Teara.govt.nz. 2009-03-02. Retrieved 2012-07-31.

Earthquake Engineering: Tools and Techniques

To measure seismic activity in the Earth's crust, a multitude of scales have been developed and each of these scales is based on measuring components like the energy released and the after-effects of the earthquake. The chapter studies the moment magnitude scale, the Richter scale, Mercalli intensity scale and the Environmental Seismic Intensity scale. There is a section on the earthquake shaking table which is a device that shakes structural models and building components in a vast range of stimulated motions, helping predict how structures react to earthquakes.

Moment Magnitude Scale

The moment magnitude scale (abbreviated as MMS; denoted as M_W or M) is used by seismologists to measure the size of earthquakes in terms of the energy released.

As with the Richter magnitude scale, an increase of one step on this logarithmic scale corresponds to a $10^{1.5} \approx 32$ times increase in the amount of energy released, and an increase of two steps corresponds to a $10^3 = 1000$ times increase in energy. Thus, an earthquake of M_W of 7.0 contains 1000 times as much energy as one of 5.0 and about 32 times that of 6.0.

The magnitude is based on the seismic moment of the earthquake, which is equal to the rigidity of the Earth multiplied by the average amount of slip on the fault and the size of the area that slipped.

The scale was developed in the 1970s to succeed the 1930s-era Richter magnitude scale (M_L). Even though the formulae are different, the new scale retains a similar continuum of magnitude values to that defined by the older one. Starting in January 2002, the MMS is officially the scale used by the United States Geological Survey to calculate and report magnitudes for all modern large earthquakes.

Popular press reports of earthquake magnitude usually fail to distinguish between magnitude scales, and are often reported as "Richter magnitudes" when the reported magnitude is a moment magnitude (or a surface-wave or body-wave magnitude). Because the scales are intended to report the same results within their applicable conditions, the confusion is minor.

Historical Context

The Richter Scale: A Former Measure of Earthquake Magnitude

In 1935, Charles Richter and Beno Gutenberg developed the local magnitude () scale (popularly known as the Richter scale) with the goal of quantifying medium-sized earthquakes (between magnitude 3.0 and 7.0) in Southern California. This scale was based on the ground motion measured by a particular type of seismometer (a Wood-Anderson seismograph) at a distance of 100 kilometres (62 mi) from the earthquake's epicenter. Because of this, there is an upper limit on the highest measurable magnitude, and all large earthquakes will tend to have a local magnitude of around 7. Further, the magnitude becomes unreliable for measurements taken at a distance of more than about 600 kilometres (370 mi) from the epicenter. Since this M_L scale was simple to use and corresponded well with the damage which was observed, it was extremely useful for engineering earthquake-resistant structures, and gained common acceptance.

The Modified Richter Scale

Although the Richter scale represented a major step forward, it was not as effective for characterizing some classes of quakes. As a result, Beno Gutenberg expanded Richter's work to consider earthquakes detected at distant locations. For such large distances the higher frequency vibrations are attenuated and seismic surface waves (Rayleigh and Love waves) are dominated by waves with a period of 20 seconds (which corresponds to a wavelength of about 60 km). Their magnitude was assigned a surface wave magnitude scale (M_S). Gutenberg also combined compressional P-waves and the transverse S-waves (which he termed "body waves") to create a body-wave magnitude scale (M_b), measured for periods between 1 and 10 seconds. Ultimately Gutenberg and Richter collaborated to produce a combined scale which was able to estimate the energy released by an earthquake in terms of Gutenberg's surface wave magnitude scale (M_S).

Correcting Weaknesses of the Modified Richter Scale

The Richter Scale, as modified, was successfully applied to characterize localities. This enabled local building codes to establish standards for buildings which were earthquake resistant. However a series of quakes were poorly handled by the modified Richter scale. This series of "great earthquakes", included faults that broke along a line of up to 1000 km. Examples include the 1952 Aleutian Fox Islands quake and the 1960 Chilean quake, both of which broke faults approaching 1000 km. The M_S scale was unable to characterize these "great earthquakes" accurately.

The difficulties with use of M_S in characterizing the quake resulted from the size of these earthquakes. Great quakes produced 20 s waves such that M_S was comparable to normal quakes, but also produced very long period waves (more than 200 s) which carried large amounts of energy. As a result, use of the modified Richter scale methodology to estimate earthquake energy was deficient at high energies.

In 1972, Keiiti Aki, a professor of Geophysics at the Massachusetts Institute of Technology, introduced elastic dislocation theory to improve understanding of the earthquake mechanism. This theory proposed that the energy release from a quake is proportional to the surface area that breaks free, the average distance that the fault is displaced, and the rigidity of the material adjacent to the fault. This is found to correlate well with the seismologic readings from long-period seismographs. Hence the moment magnitude scale (M_w) represented a major step forward in characterizing earthquakes.

Current Research

Recent research related to the moment magnitude scale focuses on:

- Timely earthquake magnitude estimates allow for early warnings of earthquakes and tsunami. Such earthquake early warning systems are operating in Japan, Mexico, Romania, Taiwan, and Turkey and are being tested in the United States, Europe, and Asia. Such systems rely on a variety of analytic methods to attain an early estimate of the moment magnitude of a quake.

- Efforts are underway to extend the moment magnitude scale accuracy for high frequencies, which are important in localizing small quakes. Earthquakes below magnitude 3 scale poorly because the earth attenuates high frequency waves near the surface, making it difficult to resolve quakes smaller than 100 meters. By use of seismographs in deep wells this attenuation can be overcome.

Definition

The symbol for the moment magnitude scale is M_w with the subscript w meaning mechanical work accomplished. The moment magnitude M_w is a dimensionless number defined by Hiroo Kanamori as

$$M_w = \frac{2}{3}\log_{10}(M_0) - 10.7,$$

where M_0 is the seismic moment in dyne·cm (10^{-7} N·m). The constant values in the equation are chosen to achieve consistency with the magnitude values produced by earlier scales, such as the Local Magnitude and the Surface Wave magnitude.

Comparative Energy Released by Two Earthquakes

As with the Richter scale, an increase of one step on this logarithmic scale corresponds to a $10^{1.5} \approx 32$ times increase in the amount of energy released, and an increase of two steps corresponds to a $10^3 = 1000$ times increase in energy. Thus, an earthquake of M_w of 7.0 contains 1000 times as much energy as one of 5.0 and about 32 times that of 6.0.

The following formula, obtained by solving the previous equation for M_0, allows one to assess the proportional difference $f_{\Delta E}$ in energy release between earthquakes of two different moment magnitudes, say m_1 and m_2 :

$$f_{\Delta E} = 10^{\frac{3}{2}(m_1 - m_2)}.$$

For example, an earthquake with a moment magnitude of 7.0 is approximately 5.62 times greater than a quake with moment magnitude 6.5.

Radiated Seismic Energy

Potential energy is stored in the crust in the form of built-up stress. During an earthquake, this stored energy is transformed and results in

- cracks and deformation in rocks

- heat

- radiated seismic energy E_s .

The seismic moment M_0 is a measure of the total amount of energy that is transformed during an earthquake. Only a small fraction of the seismic moment M_0 is converted into radiated seismic energy E_s, which is what seismographs register. Using the estimate

$$E_s = M_0 \cdot 10^{-4.8} = M_0 \cdot 1.6 \times 10^{-5},$$

Choy and Boatwright defined in 1995 the *energy magnitude*

$$M_e = \frac{2}{3} \log_{10} E_s - 2.9$$

where E_s is in N.m.

Nuclear Explosions

The energy released by nuclear weapons is traditionally expressed in terms of the energy stored in a kiloton or megaton of the conventional explosive trinitrotoluene (TNT).

A rule of thumb equivalence from seismology used in the study of nuclear proliferation asserts that a one kiloton nuclear explosion creates a seismic signal with a magnitude of approximately 4.0. This in turn leads to the equation

$$M_n = \frac{2}{3} \log_{10} \frac{m_{TNT}}{Mt} + 6,$$

where m_{TNT} is the mass of the explosive TNT that is quoted for comparison (relative to megatons Mt).

Such comparison figures are not very meaningful. As with earthquakes, during an underground explosion of a nuclear weapon, only a small fraction of the total amount of energy released ends up being radiated as seismic waves. Therefore, a seismic efficiency needs to be chosen for the bomb that is being quoted in this comparison. Using the conventional specific energy of TNT (4.184 MJ/kg), the above formula implies that about 0.5% of the bomb's energy is converted into radiated seismic energy E_s. For real underground nuclear tests, the actual seismic efficiency achieved varies significantly and depends on the site and design parameters of the test.

Comparison with Richter Scale

The moment magnitude (M_w) scale was introduced in 1979 by Caltech seismologists Thomas C. Hanks and Hiroo Kanamori to address the shortcomings of the Richter scale (detailed above) while maintaining consistency. Thus, for medium-sized earthquakes, the moment magnitude values should be similar to Richter values. That is, a magnitude 5.0 earthquake will be about a 5.0 on both scales. This scale was based on the physical properties of the earthquake, specifically the seismic moment (M_0). Unlike other scales, the moment magnitude scale does not saturate at the upper end; there is no upper limit to the possible measurable magnitudes. However, this has the side-effect that the scales diverge for smaller earthquakes.

The concept of seismic moment was introduced in 1966, but it took 13 years before the M_w scale was designed. The reason for the delay was that the necessary spectra of seismic signals had to be derived by hand at first, which required personal attention to every event. Faster computers than those available in the 1960s were necessary and seismologists had to develop methods to process earthquake signals automatically. In the mid-1970s Dziewonski started the Harvard Global Centroid Moment Tensor Catalog. After this advance, it was possible to introduce M_w and estimate it for large numbers of earthquakes.

Moment magnitude is now the most common measure for medium to large earthquake magnitudes, but breaks down for smaller quakes. For example, the United States Geological Survey does not use this scale for earthquakes with a magnitude of less than 3.5, which is the great majority of quakes.

Magnitude scales differ from earthquake intensity, which is the perceptible shaking, and local damage experienced during a quake. The shaking intensity at a given spot depends on many factors, such as soil types, soil sublayers, depth, type of displacement, and range from the epicenter (not counting the complications of building engineering and architectural factors). Rather, magnitude scales are used to estimate with one number the size of the quake.

The following table compares magnitudes towards the upper end of the Richter Scale for major Californian earthquakes.

Date	Seismic moment (dyne-cm)	Richter scale	Moment magnitude
1933-03-11	2	6.3	6.2
1940-05-19	30	6.4	7.0
1941-07-01	0.9	5.9	6.0
1942-10-21	9	6.5	6.6
1946-03-15	1	6.3	6.0
1947-04-10	7	6.2	6.5
1948-12-04	1	6.5	6.0
1952-07-21	200	7.2	7.5
1954-03-19	4	6.2	6.4

Richter Magnitude Scale

The Richter magnitude scale (also Richter scale) assigns a magnitude number to quantify the energy released by an earthquake. The Richter scale, developed in the 1930s, is a base-10 logarithmic scale, which defines magnitude as the logarithm of the ratio of the amplitude of the seismic waves to an arbitrary, minor amplitude.

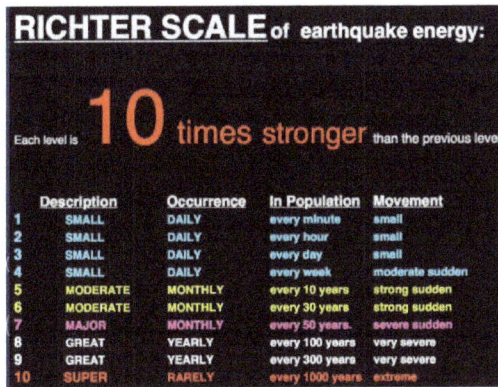

As measured with a seismometer, an earthquake that registers 5.0 on the Richter scale has a shaking amplitude 10 times that of an earthquake that registered 4.0, and thus corresponds to a release of energy 31.6 times that released by the lesser earthquake. In the United States, the Richter scale was succeeded in the 1970s by the moment magnitude scale. The moment magnitude scale is now the scale used by the United States Geological Survey to estimate magnitudes for all modern large earthquakes.

Development

In 1935, the seismologists Charles Francis Richter and Beno Gutenberg, of the California Institute of Technology, developed the (future) Richter magnitude scale, specifically for measuring earthquakes in a given area of study in California, as recorded and measured with the Wood-Anderson torsion seismograph. Originally, Richter reported mathematical values to the nearest quarter of a unit, but the values later were reported with one decimal place; the local magnitude scale compared the magnitudes of different earthquakes. Richter derived his earthquake-magnitude scale from the apparent magnitude scale used to measure the brightness of stars.

Charles Francis Richter, c. 1970

Richter established a magnitude 0 event to be an earthquake that would show a maximum, combined horizontal displacement of 1.0 μm (0.00004 in.) on a seismogram recorded with a Wood-Anderson torsion seismograph 100 km (62 mi.) from the earthquake epicenter. That fixed measure was chosen to avoid negative values for magnitude, given that the slightest earthquakes that could be recorded and located at the time were around magnitude 3.0. The Richter magnitude scale itself has no lower limit, and contemporary seismometers can register, record, and measure earthquakes with negative magnitudes.

M_L (local magnitude) was not designed to be applied to data with distances to the hypocenter of the earthquake that were greater than 600 km (373 mi.). For national and local seismological observatories, the standard magnitude scale in the 21st century is still M_L. This scale saturates at around $M_L = 7$, because the high frequency waves recorded locally have wavelengths shorter than the rupture lengths of large earthquakes.

Later, to express the size of earthquakes around the planet, Gutenberg and Richter developed a surface wave magnitude scale (M_s) and a body wave magnitude scale (M_b). These are types of waves that are recorded at teleseismic distances. The two scales were adjusted such that they were consistent with the M_L scale. That adjustment succeeded better with the scale than with the M_s scale. Each scale saturates when the earthquake is greater than magnitude 8.0.

Because of this, researchers in the 1970s developed the moment magnitude scale

(M_w). The older magnitude-scales were superseded by methods for calculating the seismic moment, from which was derived the moment magnitude scale.

About the origins of the Richter magnitude scale, C.F. Richter said:

I found a paper by Professor K. Wadati of Japan in which he compared large earthquakes by plotting the maximum ground motion against [the] distance to the epicenter. I tried a similar procedure for our stations, but the range between the largest and smallest magnitudes seemed unmanageably large. Dr. Beno Gutenberg then made the natural suggestion to plot the amplitudes logarithmically. I was lucky, because logarithmic plots are a device of the devil.

—*Charles Richter Interview, abridged from the Earthquake Information Bulletin, Vol. 12, No. 1, January-February, 1980.*

Magni-tude	Descrip-tion	Mercal-li inten-sity	Average earthquake effects	Average frequency of occurrence (estimated)
1.0–1.9	Micro	I	Microearthquakes, not felt, or felt rarely. Recorded by seismographs.	Continual/several million per year
2.0–2.9	Minor	I to II	Felt slightly by some people. No damage to buildings.	Over one million per year
3.0–3.9		III to IV	Often felt by people, but very rarely causes damage. Shaking of indoor objects can be noticeable.	Over 100,000 per year
4.0–4.9	Light	IV to VI	Noticeable shaking of indoor objects and rattling noises. Felt by most people in the affected area. Slightly felt outside. Generally causes none to minimal damage. Moderate to significant damage very unlikely. Some objects may fall off shelves or be knocked over.	10,000 to 15,000 per year
5.0–5.9	Moderate	VI to VIII	Can cause damage of varying severity to poorly constructed buildings. At most, none to slight damage to all other buildings. Felt by everyone.	1,000 to 1,500 per year
6.0–6.9	Strong	VII to X	Damage to a moderate number of well-built structures in populated areas. Earthquake-resistant structures survive with slight to moderate damage. Poorly designed structures receive moderate to severe damage. Felt in wider areas; up to hundreds of miles/kilometers from the epicenter. Strong to violent shaking in epi-central area.	100 to 150 per year

			Causes damage to most buildings, some to partially or completely collapse or receive severe damage. Well-designed structures are likely to receive damage. Felt across great distances with major damage mostly limited to 250 km from epicenter.	
7.0–7.9	Major			10 to 20 per year
8.0–8.9		VIII or greater	Major damage to buildings, structures likely to be destroyed. Will cause moderate to heavy damage to sturdy or earthquake-resistant buildings. Damaging in large areas. Felt in extremely large regions.	One per year
9.0 and greater	Great		At or near total destruction – severe damage or collapse to all buildings. Heavy damage and shaking extends to distant locations. Permanent changes in ground topography.	One per 10 to 50 years

(*Based on U.S. Geological Survey documents.*)

The intensity and death toll depend on several factors (earthquake depth, epicenter location, population density, to name a few) and can vary widely.

Minor earthquakes occur every day and hour. On the other hand, great earthquakes occur once a year, on average. The largest recorded earthquake was the Great Chilean earthquake of May 22, 1960, which had a magnitude of 9.5 on the moment magnitude scale. The larger the magnitude, the less frequent the earthquake happens.

Beyond 9.5, while extremely strong earthquakes are theoretically possible, the energies involved rapidly make such earthquakes on Earth effectively impossible without an extremely destructive source of external energy. For example, the asteroid impact that created the Chicxulub crater and caused the mass extinction that may have killed the dinosaurs has been estimated as causing a magnitude 13 earthquake, while a magnitude 15 earthquake could destroy the Earth completely. Seismologist Susan Hough has suggested that 10 may represent a very approximate upper limit, as the effect if the largest known continuous belt of faults ruptured together (along the Pacific coast of the Americas).

Energy Release Equivalents

The following table lists the approximate energy equivalents in terms of TNT explosive force – though note that the earthquake energy is released *underground* rather than overground. Most energy from an earthquake is not transmitted to and through the surface; instead, it dissipates into the crust and other subsurface structures. In contrast, a small atomic bomb blast will not, it will simply cause light shaking of indoor items, since its energy is released above ground.

Approximate magnitude	Approximate TNT equivalent for seismic energy yield	Joule equivalent	Example
0.0	15 g	63 kJ	
0.2	30 g	130 kJ	Large hand grenade
1.5	2.7 kg	11 MJ	Seismic impact of typical small construction blast
2.1	21 kg	89 MJ	West fertilizer plant explosion
3.0	480 kg	2.0 GJ	Oklahoma City bombing, 1995
3.5	2.7 metric tons	11 GJ	PEPCON fuel plant explosion, Henderson, Nevada, 1988
3.87	9.5 metric tons	40 GJ	Explosion at Chernobyl nuclear power plant, 1986
3.91	11 metric tons	46 GJ	Massive Ordnance Air Blast bomb
6.0	15 kilotons	63 TJ	Approximate yield of the Little Boy atomic bomb dropped on Hiroshima (~16 kt)
7.9	10.7 megatons	45 PJ	Tunguska event
8.35	50 megatons	210 PJ	Tsar Bomba—Largest thermonuclear weapon ever tested. Most of the energy was dissipated in the atmosphere. The seismic shock was estimated at 5.0–5.2
9.15	800 megatons	3.3 EJ	Toba eruption 75,000 years ago; among the largest known volcanic events.
13.0	100 teratons	420 ZJ	Yucatán Peninsula impact (creating Chicxulub crater) 65 Ma ago (10^8 megatons; over 4×10^{29} ergs = 400 ZJ).

Magnitude Empirical Formulae

These formulae are an alternative method to calculate Richter magnitude instead of using Richter correlation tables based on Richter standard seismic event (=0, A=0.001mm, D=100 km).

The Lillie empirical formula:

$$M_L = \log_{10} A - 2.48 + 2.76 \log_{10} \Delta$$

Where:

- A is the amplitude (maximum ground displacement) of the P-wave, in micrometers, measured at 0.8 Hz.

- Δ is the epicentral distance, in km.

For distance less than 200 km:

$$M_L = \log_{10} A + 1.6 \log_{10} D - 0.15$$

For distance between 200 km and 600 km:

$$M_L = \log_{10} A + 3.0 \log_{10} D - 3.38$$

where A is seismograph signal amplitude in mm, D distance in km.

The Bisztricsany (1958) empirical formula for epicentral distances between 4° to 160°:

$$M_L = 2.92 + 2.25 \log_{10}(\tau) - 0.001\Delta°$$

Where:

- M_L is magnitude (mainly in the range of 5 to 8)
- τ is the duration of the surface wave in seconds
- Δ is the epicentral distance in degrees.

The Tsumura empirical formula:

$$M_L = -2.53 + 2.85 \log_{10}(F - P) + 0.0014\Delta°$$

Where:

- M_L is the magnitude (mainly in the range of 3 to 5).
- $F - P$ is the total duration of oscillation in seconds.
- Δ is the epicentral distance in kilometers.

The Tsuboi, University of Tokyo, empirical formula:

$$M_L = \log_{10} A + 1.73 \log_{10} \Delta - 0.83$$

Where:

- M_L is the magnitude.
- A is the amplitude in um.
- Δ is the epicentral distance in kilometers.

Mercalli Intensity Scale

The Mercalli intensity scale is a seismic scale used for measuring the intensity of an earthquake. It measures the *effects* of an earthquake, and is distinct from the moment magnitude usually reported for an earthquake (sometimes misreported as the Richter magnitude), which is a measure of the *energy* released. The intensity of an earthquake is not entirely determined by its magnitude. It is not based on first physical principles, but is, instead, empirically based on observed effects.

The Mercalli scale quantifies the effects of an earthquake on the Earth's surface, humans, objects of nature, and man-made structures on a scale from I (not felt) to XII (total destruction). Values depend upon the distance from the earthquake, with the highest intensities being around the epicentral area. Data gathered from people who have experienced the quake are used to determine an intensity value for their location. The Italian volcanologist Giuseppe Mercalli revised the widely used simple ten-degree Rossi–Forel scale between 1884 and 1906, creating the Mercalli Intensity scale which is still used nowadays.

In 1902, the ten-degree Mercalli scale was expanded to twelve degrees by Italian physicist Adolfo Cancani. It was later completely re-written by the German geophysicist August Heinrich Sieberg and became known as the Mercalli–Cancani–Sieberg (MCS) scale.

The Mercalli–Cancani–Sieberg scale was later modified and published in English by Harry O. Wood and Frank Neumann in 1931 as the Mercalli–Wood–Neumann (MWN) scale. It was later improved by Charles Richter, the father of the Richter magnitude scale.

The scale is known today as the Modified Mercalli scale (MM) or Modified Mercalli Intensity scale (MMI).

Modified Mercalli Intensity Scale

The lower degrees of the Modified Mercalli Intensity scale generally deal with the manner in which the earthquake is felt by people. The higher numbers of the scale are based on observed structural damage.

The large table gives Modified Mercalli scale intensities that are typically observed at locations near the epicenter of the earthquake.

I. Not felt	Not felt except by a very few under especially favorable conditions.
II. Weak	Felt only by a few people at rest, especially on upper floors of buildings.
III. Weak	Felt quite noticeably by people indoors, especially on upper floors of buildings. Many people do not recognize it as an earthquake. Standing motor cars may rock slightly. Vibrations similar to the passing of a truck. Duration estimated.
IV. Light	Felt indoors by many, outdoors by few during the day. At night, some awakened. Dishes, windows, doors disturbed; walls make cracking sound. Sensation like heavy truck striking building. Standing motor cars rocked noticeably.
V. Moderate	Felt by nearly everyone; many awakened. Some dishes, windows broken. Unstable objects overturned. Pendulum clocks may stop.
VI. Strong	Felt by all, many frightened. Some heavy furniture moved; a few instances of fallen plaster. Damage slight.
VII. Very strong	Damage negligible in buildings of good design and construction; slight to moderate in well-built ordinary structures; considerable damage in poorly built or badly designed structures; some chimneys broken.

VIII. Severe	Damage slight in specially designed structures; considerable damage in ordinary substantial buildings with partial collapse. Damage great in poorly built structures. Fall of chimneys, factory stacks, columns, monuments, walls. Heavy furniture overturned.
IX. Violent	Damage considerable in specially designed structures; well-designed frame structures thrown out of plumb. Damage great in substantial buildings, with partial collapse. Buildings shifted off foundations.
X. Extreme	Some well-built wooden structures destroyed; most masonry and frame structures destroyed with foundations. Rails bent.
XI. Extreme	Few, if any, (masonry) structures remain standing. Bridges destroyed. Broad fissures in ground. Underground pipe lines completely out of service. Earth slumps and land slips in soft ground. Rails bent greatly.
XII. Extreme	Damage total. Waves seen on ground surfaces. Lines of sight and level distorted. Objects thrown upward into the air.

Correlation with Magnitude

Magnitude	Typical Maximum Modified Mercalli Intensity
Under 2.0	I
2.0 – 2.9	II – III
3.0 – 3.9	III – IV
4.0 – 4.9	IV – V
5.0 – 5.9	V – VI
6.0 – 6.9	VI – VII
7.0 – 7.9	VII – VIII
8.0 or higher	VIII or higher

The correlation between magnitude and intensity is far from total, depending upon several factors including the depth of the earthquake, terrain, population density, and damage. For example, on May 19, 2011, an earthquake of magnitude 0.7 in Central California, United States, 4 km deep was classified as of intensity III by the United States Geological Survey (USGS) over 100 miles (160 km) away from the epicenter (and II intensity almost 300 miles (480 km) from the epicenter), while a 4.5 magnitude quake in Salta, Argentina, 164 km deep was of intensity I.

The small table is a rough guide to the degrees of the Modified Mercalli Intensity scale. The colors and descriptive names shown here differ from those used on certain shake maps in other articles.

Correlation with Physical Quantities

The Mercalli scale is not defined in terms of more rigorous, objectively quantifiable measurements such as shake amplitude, shake frequency, peak velocity, or peak acceleration.

Human-perceived shaking and building damages are best correlated with peak acceleration for lower-intensity events, and with peak velocity for higher-intensity events.

Comparison to the Moment Magnitude Scale

The effects of any one earthquake can vary greatly from place to place, so there may be
many Mercalli intensity values measured for the same earthquake. These values can be
best displayed using a contoured map of equal intensity, known as an isoseismal map.
However, each earthquake has only one magnitude.

Earthquake Shaking Table

There are several different experimental techniques that can be used to test the response of structures to verify their seismic performance, one of which is the use of
an earthquake shaking table (a shaking table, or simply shake table). This is a device
for shaking structural models or building components with a wide range of simulated
ground motions, including reproductions of recorded earthquakes time-histories.

While modern tables typically consist of a rectangular platform that is driven in up
to six degrees of freedom (DOF) by servo-hydraulic or other types of actuators, the
earliest shake table, invented at the University of Tokyo in 1893 to categorize types of
building construction, ran on a simple wheel mechanism. Test specimens are fixed to
the platform and shaken, often to the point of failure. Using video records and data
from transducers, it is possible to interpret the dynamic behaviour of the specimen.
Earthquake shaking tables are used extensively in seismic research, as they provide the
means to excite structures in such a way that they are subjected to conditions represen-
tative of true earthquake ground motions.

They are also used in other fields of engineering to test and qualify vehicles and components of vehicles that must respect heavy vibration requirements and standards (i.e.
aerospace, military standards etc.).

A World List of Shaking Tables

Below is an attempt to create a definitive list of shaking tables around the world that are
used for seismic testing. The list is almost certainly not complete and not all the data
has been verified by the shaking table owners. If you operate a shaking table please help
by correcting and updating and adding to this list.

This list was originally based on information from the following documents: Exper-
imental Facilities for Earthquake Engineering Simulation Worldwide, Directory of
International Earthquake Engineering Research Facilities and papers by Chowdhury,
Duarte, Kamimura and Nakashima and EERI.

Shaking Tables around the world

Region	Country	State	Location	Size (m)	Payload (metric tonnes)	Degrees of Freedom	X Horiz Disp (mm)	Y Horiz Disp (mm)	Z Vert Disp (mm)	X Horiz Vel (mm/s)	Y Horiz vel (mm/s)	Z Vert vel (mm/s)	X Horiz accel (m/s2)	Y Horiz accel (m/s2)	Z Vert accel (m/s2)	Max Freq (Hz)	Details checked
Africa	Algeria	-	CGS Laboratory, Alger	6.1 x 6.1	60	6	±150	±250	±100	±1100	±1100	±1000	±10	±10	±8	100	30/6/2010
Africa	South Africa	-	University of Witwatersrand, Johannesburg	4 x 4	10	1	±750	n/a	n/a	±1000	n/a	n/a	±10	n/a	n/a	40	17/7/2009
Asia	China	-	China Academy of Building Research, Beijing	6.1 x 6.1	60	6	±150	±250	±100	±1000	±1200	±800	±15	±10	±8	50	?
Asia	China	-	Guangzhou University	3 x 3	20	6	±100	±100	±50	±1000	±1000	±1000	±26	±26	±50	50	10/7/2008
Asia	China	-	Nanjing University of Technology	3 x 5	15	3	±120	±120	±120	±500	±500	±500	±10	±10	±10	50	?
Asia	China	-	Tongji University, Shanghai	4 x 4	25	6	±100	±50	±50	±1000	±600	±600	±40	±20	±50	50	?
Asia	China	-	Xi'a University of Architecture & Technology	2 x 2	?	?	?	?	?	?	?	?	?	?	?	?	?
Asia	Singapore	-	Nanyang Technological University	3 x 3	10	1	±120	n/a	n/a	±650	n/a	n/a	±20	n/a	n/a	50	23/7/2008

Shaking Tables around the world

Region	Country	State	Location	Size (m)	Payload (metric tonnes)	Degrees of Freedom	X Horiz Disp (mm)	Y Horiz Disp (mm)	Z Vert Disp (mm)	X Horiz Vel (mm/s)	Y Horiz vel (mm/s)	Z Vert vel (mm/s)	X Horiz accel (m/s2)	Y Horiz accel (m/s2)	Z Vert accel (m/s2)	Max Freq (Hz)	Details checked
Asia	Hong Kong	-	City University of Hong Kong	?	?	?	?	?	?	?	?	?	?	?	?	?	?
Asia	Hong Kong	-	Hong Kong Polytechnic University	3 x 3	10	?	?	?	?	?	?	?	10	?	?	?	?
Asia	India	Delhi	Jamia Millia Islamia, New Delhi	5 x 5	20	1	±500	n/a	n/a	?	?	?	±20	n/a	n/a	100	?
Asia	India	Assam	IIT Guwahati	2.5 x 2.5	5	1	±500	n/a	n/a	?	?	?	±20	n/a	n/a	100	?
Asia	India	Karnataka	CPRI Bangalore, Karnataka	3 x 3	10	6	?	?	?	?	?	?	?	?	?	?	?
Asia	India	Karnataka	IISc, Bangalore	1 x 1	0.5	6	±220	±220	±100	±570	±570	±570	±30	±30	±20	50	23/7/2008
Asia	India	Tamil Nadu	SERC, Chennai (1 of 3), Tamil Nadu	4 x 4	30	3	?	?	?	?	?	?	?	?	?	?	?
Asia	India	Tamil Nadu	SERC, Chennai (2 of 3), Tamil Nadu	2 x 2	5	3	?	?	?	?	?	?	?	?	?	?	?
Asia	India	Tamil Nadu	SERC, Chennai (3 of 3), Tamil Nadu	3 x 3	10	6	?	?	?	?	?	?	?	?	?	?	?

Shaking Tables around the world

Region	Country	State	Location	Size (m)	Payload (metric tonnes)	Degrees of Freedom	X Horiz Disp (mm)	Y Horiz Disp (mm)	Z Vert Disp (mm)	X Horiz Vel (mm/s)	Y Horiz vel (mm/s)	Z Vert vel (mm/s)	X Horiz accel (m/s2)	Y Horiz accel (m/s2)	Z Vert accel (m/s2)	Max Freq (Hz)	Details checked
Asia	India	Tamil Nadu	Indira Gandhi Centre for Atomic Research(IGCAR), Chennai, Tamil Nadu	3 x 3	10	6	±100	±100	±100	300	300	?	±14.715	±14.715	9.81	100	?
Asia	India	Uttarakhand	IIT Roorkee, Uttarakhand	3.5 x 3.5	20	2	?	n/a	?	?	n/a	?	?	n/a	?	?	?
Asia	India	Uttar Pradesh	IIT Kanpur, Uttar Pradesh	1.2 x 1.8	4	1	±75	n/a	n/a	±1500	n/a	n/a	±50	n/a	n/a	50	25/6/2009
Asia	India	West Bengal	Bengal Engineering and Science University, Shibpur, West Bengal	1.5 x 1.5	?	1	?	n/a	n/a	?	n/a	n/a	?	n/a	n/a	?	19/11/2009
Asia	Iran	-	International Institute of Earthquake Engineering and Seismology(IIEES)	1.4 x 1.2	2	1	±35	n/a	n/a	?	n/a	n/a	±40	n/a	n/a	?	?

Shaking Tables around the world

Region	Country	State	Location	Size (m)	Payload (metric tonnes)	Degrees of Freedom	X Horiz Disp (mm)	Y Horiz Disp (mm)	Z Vert Disp (mm)	X Horiz Vel (mm/s)	Y Horiz vel (mm/s)	Z Vert vel (mm/s)	X Horiz accel (m/s2)	Y Horiz accel (m/s2)	Z Vert accel (m/s2)	Max Freq (Hz)	Details checked
Asia	Iran	-	Iran University of Science & Technology (IUST)	2 x 0.5	5	1	±60	n/a	n/a	?	n/a	n/a	±6.5	n/a	n/a	?	?
Asia	Iran	-	Sharif University of Technology (SUT))	4 x 4	30	3	±125	±200	?	±500	±800	?	±50	±40	?	50	7/19/2011
Asia	Japan	-	Aichi Institute of Technology	11 x 6	136	1	±150	?	?	±1000	?	?	±10	?	?	50	?
Asia	Japan	-	Building Research Institute	3 x 4	20	3	?	?	?	?	?	?	?	?	?	?	?
Asia	Japan	-	Central Research Institute of Electric Power Industry	5 x 5	60	1	±500	n/a	n/a	±1500	n/a	n/a	±10	n/a	n/a	30	12/3/2008
Asia	Japan	-	NIED 'E-Defence' Laboratory, Miki City	20 x 15	1200	6	±1000	±1000	±500	±2000	±2000	±700	±9	±9	±15	50	3/3/2008
Asia	Japan	-	Fujita Corporation	4 x 4	25	1	±500	n/a	n/a	±1500	n/a	n/a	±10	n/a	n/a	50	?
Asia	Japan	-	Hazama Corp Ltd.	6 x 4	80	3	±300	±150	±100	±1150	?	?	±20	±3	±2	50	?

Shaking Tables around the world

Region	Country	State	Location	Size (m)	Payload (metric tonnes)	Degrees of Freedom	X Horiz Disp (mm)	Y Horiz Disp (mm)	Z Vert Disp (mm)	X Horiz Vel (mm/s)	Y Horiz vel (mm/s)	Z Vert vel (mm/s)	X Horiz accel (m/s2)	Y Horiz accel (m/s2)	Z Vert accel (m/s2)	Max Freq (Hz)	Details checked
Asia	Japan	-	Hitachi Engineering Corp	4 x 4	20	1	±150	?	?	±750	?	?	±20	?	?	30	?
Asia	Japan	-	Ishikawajima Harima Heavy Ind Corp.	4.5 x 4.5	35	6	±100	±100	±67	±750	±750	±500	±15	±15	±10	50	?
Asia	Japan	-	JDC Corp.	?	20	6	±300	?	?	±1000	?	?	±10	?	?	50	?
Asia	Japan	-	Kawasaki Heavy Industries Corp.	?	30	3	±75	?	?	±400	?	?	±10	?	?	50	?
Asia	Japan	-	Kajima Corp. Ltd. (1 of 2)	5 x 5	50	6	±200	±200	±100	±1000	±1000	±500	±20	±20	±20	60	3/3/2008
Asia	Japan	-	Kajima Corp. Ltd. (2 of 2)	4 x 4	20	2	±150	n/a	±75	±1140	n/a	±4450	±20	n/a	±10	50	?
Asia	Japan	-	Kumagai-Gumi Corp Ltd	5 x 5	64	6	±80	±260	±50	±600	±1500	±500	±30	±10	±10	70	?
Asia	Japan	-	Kyoto University	5 x 3	14	6	±300	?	?	±1500	?	?	±10	?	?	50	?
Asia	Japan	-	Kyoto University Disaster Prevention Research Centre	3.5 diameter	?	6	?	?	?	?	?	?	?	?	?	?	?
Asia	Japan	-	Ministry of Construction	6 x 8	100	2	±75	n/a	?	±600	n/a	?	±7	n/a	?	30	?

Shaking Tables around the world

Region	Country	State	Location	Size (m)	Payload (metric tonnes)	Degrees of Freedom	X Horiz Disp (mm)	Y Horiz Disp (mm)	Z Vert Disp (mm)	X Horiz Vel (mm/s)	Y Horiz vel (mm/s)	Z Vert vel (mm/s)	X Horiz accel (m/s2)	Y Horiz accel (m/s2)	Z Vert accel (m/s2)	Max Freq (Hz)	Details checked
Asia	Japan	-	Mitsubishi Electric Corp.	?	40	2	±100	?	?	±700	?	?	±25	?	?	30	?
Asia	Japan	-	Mitsubishi Heavy Industries Corp.	?	100	3	±50	?	?	±1500	?	?	±20	?	?	50	?
Asia	Japan	-	National Research Institute of Agriculture Engineering	6 x 4	45	3	±150	?	?	±750	?	?	±10	?	?	40	?
Asia	Japan	-	NIED (Nat. Inst. for Disaster Prevention) (1 of 2)	6 x 6	1100	3	±1000	?	?	±2000	?	?	±10	?	?	15	?
Asia	Japan	-	NIED (Nat. Inst. for Disaster Prevention) (2 of 2)	12 x 12	500	1	±220	?	?	±900	?	?	±10	?	?	50	?
Asia	Japan	-	Nishimatsu Construction Corp	5.5 x 5.5	65	6	±500	?	?	±1500	?	?	±20	?	?	50	?
Asia	Japan	-	Nuclear Power Engineering Corporation	15 x 15	908	2	±200	n/a	±100	±750	n/a	±375	±18	n/a	±9	30	?

Shaking Tables around the world

Region	Country	State	Location	Size (m)	Payload (metric tonnes)	Degrees of Freedom	X Horiz Disp (mm)	Y Horiz Disp (mm)	Z Vert Disp (mm)	X Horiz Vel (mm/s)	Y Horiz vel (mm/s)	Z Vert vel (mm/s)	X Horiz accel (m/s2)	Y Horiz accel (m/s2)	Z Vert accel (m/s2)	Max Freq (Hz)	Details checked
Asia	Japan	-	NYK Corporation	2.6 x 2.6	20	6	±200	?	?	±600	?	?	±20	?	?	80	?
Asia	Japan	-	Obayashi-Gumi Corporation	5 x 5	46	3	±600	?	?	±2000	?	?	±30	?	?	50	?
Asia	Japan	-	Okumura Corp.	?	60	6	±125	?	?	±1000	?	?	±30	?	?	70	?
Asia	Japan	-	Penta-Ocean Construction Co. Ltd. (1 of 2)	?	60	6	±300	?	?	±1000	?	?	±5	?	?	70	?
Asia	Japan	-	Penta-Ocean Construction Co. Ltd. (2 of 2)	?	60	6	±200	?	?	±750	?	?	±10	?	?	70	?
Asia	Japan	-	Port and Airport Research Institute	3.4 x 3.4	55	2	±200	n/a	±100	±750	±500	n/a	±8	±15	n/a	50	?
Asia	Japan	-	Public Works Research Institute (PWRI)	8 x 8	300	6	±600	±600	±300	±2000	±2000	±1000	±20	±20	±10	50	10/5/2008
Asia	Japan	-	Sanryo Heavy Industries Corporation	6 x 6	90.7	3	?	?	?	?	?	?	?	?	?	?	?

Shaking Tables around the world

Region	Country	State	Location	Size (m)	Payload (metric tonnes)	Degrees of Freedom	X Horiz Disp (mm)	Y Horiz Disp (mm)	Z Vert Disp (mm)	X Horiz Vel (mm/s)	Y Horiz vel (mm/s)	Z Vert vel (mm/s)	X Horiz accel (m/s2)	Y Horiz accel (m/s2)	Z Vert accel (m/s2)	Max Freq (Hz)	Details checked
Asia	Japan	-	Shimizu Corporation	4 x 4	20	3	±200	?	?	±1000	?	?	±10	?	?	50	?
Asia	Japan	-	Taisei Corp Ltd	4 x 4	20	2	±200	?	?	±450	?	?	±10	?	?	50	?
Asia	Japan	-	Tobishima Corp Ltd	4 x 4	20	2	±75	?	?	?	?	?	±15	?	?	30	?
Asia	Japan	-	Tobishima Corp Ltd	?	20	6	±200	?	?	±800	?	?	±10	?	?	30	?
Asia	Japan	-	Tokyu Const. Corp.	4 x 4	30	6	±500	±200	±100	±1500	±1000	?	±10	?	?	30	?
Asia	Japan	-	Toshiba Electric Co.	5 x 5	20	2	±750	n/a	±380	±400	n/a	±250	±10	n/a	±7	30	?
Asia	South Korea	-	Hyundai Institute of Construction Technology Development (1 of 2)	2 x 2	5	2	±75	±75	n/a	±500	±500	n/a	±10	±10	n/a	?	5/2/2009
Asia	South Korea	-	Hyundai Institute of Construction Technology Development (2 of 2)	5 x 3	30	1	±100	n/a	n/a	±500	n/a	n/a	±10	n/a	n/a	?	5/2/2009
Asia	South Korea	-	Korea Institute of Machinery and Metals, Changwon	4 x 4	30	6	±200	±200	±134	±750	±750	±500	±15	±15	±10	50	?

Shaking Tables around the world

Region	Country	State	Location	Size (m)	Payload (metric tonnes)	Degrees of Freedom	X Horiz Disp (mm)	Y Horiz Disp (mm)	Z Vert Disp (mm)	X Horiz Vel (mm/s)	Y Horiz vel (mm/s)	Z Vert vel (mm/s)	X Horiz accel (m/s2)	Y Horiz accel (m/s2)	Z Vert accel (m/s2)	Max Freq (Hz)	Details checked
Asia	Korea	-	Pusan National University (1 of 3)	5 x 5	300	2	±300	±200	n/a	±1000	±1000	n/a	±20	±20	n/a	60	24/7/2008
Asia	Korea	-	Pusan National University (2 of 3)	5 x 5	600	2	±300	±200	n/a	±1000	±1000	n/a	±30	±30	n/a	60	24/7/2008
Asia	Korea	-	Pusan National University (3 of 3)	4 x 4	300	6	±300	±200	±150	±1500	±1500	±1000	±20	±20	±40	60	24/7/2008
Asia	Malaysia	-	Sabah University	1.5 x 1.5	?	?	?	?	?	?	?	?	?	?	?	?	?
Asia	Taiwan	-	National Center for Research in Earthquake Engineering	5 x 5	50	6	±250	±100	±100	±1000	±600	±500	±10	±30	±10	50	12/3/2008
Oceania	Australia	Victoria	University of Melbourne	2 x 2	3	?	?	?	?	?	?	?	?	?	?	?	?
Oceania	Australia	New South Wales	University of Technology, Sydney	3 x 3	10	1	?	?	?	?	?	?	?	?	?	50	15/12/2009
North America	Canada	Ontario	Royal Military College of Canada, Kingston	2.7 x 2.7	5	1	?	?	?	?	?	?	?	?	?	?	?

Shaking Tables around the world

Region	Country	State	Location	Size (m)	Payload (metric tonnes)	Degrees of Freedom	X Horiz Disp (mm)	Y Horiz Disp (mm)	Z Vert Disp (mm)	X Horiz Vel (mm/s)	Y Horiz vel (mm/s)	Z Vert vel (mm/s)	X Horiz accel (m/s2)	Y Horiz accel (m/s2)	Z Vert accel (m/s2)	Max Freq (Hz)	Details checked
North America	Canada	British Columbia	University of British Columbia (EERF Lab), Vancouver	4 x 4	30	6	?	?	?	?	?	?	?	?	?	?	?
North America	Canada	Québec	University of Sherbrooke	? x ?	?	?	?	?	?	?	?	?	?	?	?	?	?
Europe	France	-	Commissariat à l'Energie Atomique et aux Energies Alternatives (CEA), AZALEE (1 of 3)	6 x 6	100	6	±125	±125	±100	±700	±700	±700	±10	±10	±25	50	27/01/2011
Europe	France	-	Commissariat à l'Energie Atomique et aux Energies Alternatives (CEA), VESUVE (2 of 3)	4 x 3	20	1	±125	n/a	n/a	±1000	n/a	n/a	±10	n/a	n/a	50	27/01/2011

Shaking Tables around the world

Region	Country	State	Location	Size (m)	Payload (metric tonnes)	Degrees of Freedom	X Horiz Disp (mm)	Y Horiz Disp (mm)	Z Vert Disp (mm)	X Horiz Vel (mm/s)	Y Horiz vel (mm/s)	Z Vert vel (mm/s)	X Horiz accel (m/s2)	Y Horiz accel (m/s2)	Z Vert accel (m/s2)	Max Freq (Hz)	Details checked
Europe	France	-	Commissariat à l'Energie Atomique et aux Energies Alternatives (CEA), TOURNESOL (3 of 3)	2 x 2	10	3	±125	n/a	±100	±2000	n/a	±1300	±10	n/a	10	50	27/01/2011
Europe	France	-	FCBA Institut technologique, Bordeaux	6 x 6	10	1	±125	n/a	n/a	±800	n/a	n/a	±20	n/a	n/a	50	02/2011
Europe	Greece	-	Aristotle University of Thessaloniki	1.2 x 1.2	15	2	±50	n/a	±50	?	n/a	?	?	n/a	?	30	?
Europe	Greece	-	National Technical University of Athens	4 x 4	10	6	±100	±100	±100	±1000	±1000	±1000	±20	±20	±40	100	12/3/2008
Europe	Italy	-	Ansaldo Nucleare, Genoa	3.5 x 3.5	6	3	±70	?	?	±860	?	?	?	?	?	60	?

Shaking Tables around the world

Region	Country	State	Location	Size (m)	Payload (metric tonnes)	Degrees of Freedom	X Horiz Disp (mm)	Y Horiz Disp (mm)	Z Vert Disp (mm)	X Horiz Vel (mm/s)	Y Horiz vel (mm/s)	Z Vert vel (mm/s)	X Horiz accel (m/s2)	Y Horiz accel (m/s2)	Z Vert accel (m/s2)	Max Freq (Hz)	Details checked
Europe	Italy	-	Department of Structures for Engineering and Architecture, University of Naples - Shake Table	4 x 4	20	2	±250	±250	n/a	±1000	±1000	n/a	10	10	n/a	50	-
Europe	Italy	-	ENEA (Casaccia R. C.) - System 1 Shake Table (1 of 2)	4 x 4	30	6	±125	±125	±125	±500	±500	±500	±30	±30	±30	50	13/6/2015
Europe	Italy	-	ENEA (Casaccia R. C.) - System 2 Shake Table (2 of 2)	2 x 2	5	6	±150	±150	±150	±1000	±1000	±1000	±50	±50	±50	100	13/6/2015
Europe	Italy	-	CESI S.p.A., Static & Dynamic Testing Laboratories, Seriate (BG)	4 x 4	30	6	±100	±100	±100	±440	±440	±440	±50	±50	?	120	13/6/2015

Shaking Tables around the world

Region	Country	State	Location	Size (m)	Payload (metric tonnes)	Degrees of Freedom	X Horiz Disp (mm)	Y Horiz Disp (mm)	Z Vert Disp (mm)	X Horiz Vel (mm/s)	Y Horiz vel (mm/s)	Z Vert vel (mm/s)	X Horiz accel (m/s2)	Y Horiz accel (m/s2)	Z Vert accel (m/s2)	Max Freq (Hz)	Details checked
Europe	Italy	-	European Centre for Training & Research in Earthquake Engineering (EUCENTRE) - Shake Table	5.6 x 7	140	1	±500	-	-	±2200	-	-	59	-	-	50	-
Europe	Italy	-	European Centre for Training & Research in Earthquake Engineering (EUCENTRE) - Bearing Testing System	1.6 x 4.4	5000 (vertical load)	5	±500	±265	140	±2200	±600	±250	18	-	-	20	-
Europe	Italy	-	European Centre for Training & Research in Earthquake Engineering (EUCENTRE) - Damper Testing System	1.5 x 2	50	1	±250	-	-	±1180	-	-	10	-	-	20	-

Shaking Tables around the world

Region	Country	State	Location	Size (m)	Payload (metric tonnes)	Degrees of Freedom	X Horiz Disp (mm)	Y Horiz Disp (mm)	Z Vert Disp (mm)	X Horiz Vel (mm/s)	Y Horiz vel (mm/s)	Z Vert vel (mm/s)	X Horiz accel (m/s2)	Y Horiz accel (m/s2)	Z Vert accel (m/s2)	Max Freq (Hz)	Details checked
Europe	Macedonia	-	Institute of Earthquake Engineering and Engineering Seismology (IZIIS), University of SS. Cyril and Methodius in Skopje (1 of 2)	5 x 5	40	5	n/a	±125	±60	n/a	±1000	±500	n/a	±30	±15	80	20/9/2011
Europe	Macedonia	-	Institute of Earthquake Engineering and Engineering Seismology (IZIIS), University of SS. Cyril and Methodius in Skopje (2 of 2)	1.4 x 2.5	8	1	±100	n/a	n/a	?	n/a	n/a	±20	n/a	n/a	80	7/7/2008
Europe	The Netherlands	-	European Space Agency (ESA) ES-TEC Test Centre, Noordwijk	5.5 x 5.5	22.5	6	±70	±70	±70	±800	±800	±800	±50	±50	±50	2-100	23/7/2008

Shaking Tables around the world

Region	Country	State	Location	Size (m)	Payload (metric tonnes)	Degrees of Freedom	X Horiz Disp (mm)	Y Horiz Disp (mm)	Z Vert Disp (mm)	X Horiz Vel (mm/s)	Y Horiz vel (mm/s)	Z Vert vel (mm/s)	X Horiz accel (m/s2)	Y Horiz accel (m/s2)	Z Vert accel (m/s2)	Max Freq (Hz)	Details checked
Europe	Portugal	-	Laboratorio Nacional de Engenharia Civil (LNEC), Lisbon	5.6 x 5.6	40	3	±175	±175	±175	±200	±200	±200	±18	±11	±6	20	?
Europe	Russia	-	Hydroproject Research Institute	5 x 5	50	3	±70	±70	±40	±600	?	?	±20	±20	?	40	?
Europe	Spain	-	CEDEX, Madrid	3 x 3	10	6	±100	±100	±50	?	?	?	±10	±10	±20	60	13/6/2015
Europe	Turkey	-	Istanbul Technical University, ITU-STEELab (Structural and Earthquake Engineering Laboratory), Istanbul	2.35 x 2.35	30	1	±325	-	-	±1250	-	-	±20	-	-	8	
Europe	Turkey	-	Bogazici University, Istanbul (1 of 2)	3 x 3	10	1	±120	n/a	n/a	±650	n/a	n/a	±20	n/a	n/a	50	3/3/2008
Europe	Turkey	-	Bogazici University, Istanbul (2 of 2)	0.7 x 0.7	0.1	3	±120	±120	±120	±1200	±1200	±1200	±100	±100	±100	40	3/3/2008

Shaking Tables around the world

Region	Country	State	Location	Size (m)	Payload (metric tonnes)	Degrees of Freedom	X Horiz Disp (mm)	Y Horiz Disp (mm)	Z Vert Disp (mm)	X Horiz Vel (mm/s)	Y Horiz vel (mm/s)	Z Vert vel (mm/s)	X Horiz accel (m/s2)	Y Horiz accel (m/s2)	Z Vert accel (m/s2)	Max Freq (Hz)	Details checked
Europe	Turkey	-	Middle East Technical University, Ankara	2 x 1	?	?	?	?	?	?	?	?	?	?	?	?	?
Europe	Bulgaria	Sofia	Lab for numerical and experimental dynamic modelling, UACEG, Sofia	0.5 x 0.36	0.01	1	±30	n/a	n/a	±300	n/a	n/a	±5	n/a	n/a	5	18/02/2016
Europe	UK	-	University of Bristol (EERC)	3 x 3	17	6	±150	±150	±150	±1100	±1100	±1100	±60	±60	±60	100	29/2/2008
Asia	Pakistan	-	Earthquake Engineering Center, University of Engineering & Technology Peshawar	6.0 x 6.0	60	6	±300	±300	±300	±1100	±1100	±1100	±14.7	±14.7	±14.7	50	
North America	Mexico	Mexico D. F.	Universidad Nacional Autónoma de México (UNAM), México City	4 x 4	20	5	±150	±150	±75	±1100	±1100	±450	±10	±10	±10	60	10/9/2008

Shaking Tables around the world

Region	Country	State	Location	Size (m)	Payload (metric tonnes)	Degrees of Freedom	X Horiz Disp (mm)	Y Horiz Disp (mm)	Z Vert Disp (mm)	X Horiz Vel (mm/s)	Y Horiz vel (mm/s)	Z Vert vel (mm/s)	X Horiz accel (m/s2)	Y Horiz accel (m/s2)	Z Vert accel (m/s2)	Max Freq (Hz)	Details checked
North America	USA	Colorado	ANCO Engineers, Inc Boulder, Colorado	3 x 3	10	3	±200	±200	±200	±2000	±2000	±2000	±30	±30	±30	40	16/10/2012
North America	USA	Alabama	NASA	3 x 4.5	1	6	±2440	?	?	±100	?	?	?	?	?	?	?
North America	USA	Ohio	NASA Mechanical Vibration Facility (MVF), Glenn's Space Power Facility, Sandusky	6.7 diameter	34	6	±15	±15	?	±338	±338	?	?	?	?	150	23/6/2015
North America	USA	Alabama	Wyle Laboratories (1 of 3)	6.1 x 5.5	27	2	±152	?	?	±890	?	?	±60	?	?	100	16/10/2012
North America	USA	Alabama	Wyle Laboratories (2 of 3)	2.7 x 2.7	4.5	3	±250	±250	±250	±1120	±1120	±1120	±45	±45	±45	100	16/10/2012
North America	USA	Alabama	Wyle Laboratories (3 of 3)	2.4 x 2.4	4.5	2	±305	n/a	±228	±1168	n/a	±838	±70	n/a	±80	70	16/10/2012
North America	USA	North Carolina	Duke University	1.2 x 1.2	5	1	±75	n/a	n/a	±500	n/a	n/a	±50	n/a	n/a	60	15/4/2008
North America	USA	California	University of California at Berkeley	6.1 x 6.1	85	6	±127	±127	±51	±762	±762	±254	±30	±30	±20	40	30/4/2008

Shaking Tables around the world

Region	Country	State	Location	Size (m)	Payload (metric tonnes)	Degrees of Freedom	X Horiz Disp (mm)	Y Horiz Disp (mm)	Z Vert Disp (mm)	X Horiz Vel (mm/s)	Y Horiz vel (mm/s)	Z Vert vel (mm/s)	X Horiz accel (m/s2)	Y Horiz accel (m/s2)	Z Vert accel (m/s2)	Max Freq (Hz)	Details checked
North America	USA	California	California State University, Fresno	2.4 x 2.0	?	1	±125	n/a	n/a	?	n/a	n/a	?	n/a	n/a	?	19/11/2009
North America	USA	California	University of California at San Diego	12.2 x 7.6	2000	1	±750	n/a	n/a	±1800	n/a	n/a	±10	n/a	n/a	20	7/7/2008
North America	USA	Connecticut	University of Connecticut	1.5 x 1.5	1	1	±150	n/a	n/a	?	n/a	n/a	±20	n/a	n/a	50	8/5/2008
North America	USA	Indiana	Purdue University	3.7 x 3.7	5	1	±50	n/a	n/a	305	n/a	n/a	19.6	n/a	n/a	50	2016-06-03
North America	USA	New York	University at Buffalo (State University of New York) (2 identical tables of 3)	3.6 x 3.6	50	6	±150	±150	±75	±1250	±1250	±500	±12	±12	±12	100	12/3/2008
North America	USA	New York	University at Buffalo (State University of New York) (3 of 3)	3.7 x 3.7	50	5	±150	n/a	±75	±762	n/a	±500	±12	n/a	±23	50	12/3/2008
North America	USA	New York	Rensselaer Polytechnic Institute	1.7 x 2.6	5	1	±130	n/a	n/a	±270	n/a	n/a	±20	n/a	n/a	50	8/5/2008

Shaking Tables around the world

Region	Country	State	Location	Size (m)	Payload (metric tonnes)	Degrees of Freedom	X Horiz Disp (mm)	Y Horiz Disp (mm)	Z Vert Disp (mm)	X Horiz Vel (mm/s)	Y Horiz vel (mm/s)	Z Vert vel (mm/s)	X Horiz accel (m/s2)	Y Horiz accel (m/s2)	Z Vert accel (m/s2)	Max Freq (Hz)	Details checked
North America	USA	Nevada	University of Nevada at Reno (3 identical biaxial tables)	4.3 x 4.5	45	2	±300	±300	n/a	±1270	±1270	n/a	±20	±20	n/a	50	8/2/2010
North America	USA	Nevada	University of Nevada at Reno (6 axis table)	2.75 x 2.75	50	6	±75	±300	±100	?	?	?	±20	±40	±10	50	13/3/2008
North America	USA	Nevada	Dynamic Certification Laboratories	2.0 diameter	4.5	6	±140	±120	±150	±1000	±1000	±1200	±98	±98	±108	100	17/7/2012
North America	USA	Texas	Rice University	0.465 m²	7	1	±75	n/a	n/a	±1400	n/a	n/a	±20	n/a	n/a	50	8/5/2008
North America	USA	Illinois	Civil Engineering Research Lab U.S. Army, Champaign	3.6 x 3.6	45	3	±300	?	?	±1300	?	?	±20	?	?	250	?
North America	USA	Virginia	The George Washington University - Virginia Campus	3 x 3	?	6	?	?	?	?	?	?	?	?	?	?	?

Notes: This list is restricted to shaking tables bigger than 2m by 2m or with a capacity of more than 4 tonnes (i.e. tables suitable for seismic testing).

Environmental Seismic Intensity Scale

The Environmental Seismic Intensity scale (ESI 2007) is a seismic scale used for measuring the intensity of an earthquake on the basis of the effects of the earthquake on the natural environment (Earthquake Environmental Effects).

The international effort to develop a new scale of macroseismic intensity that would focus exclusively on environmental effects of earthquakes began in the early 1990s and was sponsored by the International Union for Quaternary Research (INQUA). After the final draft of the scale was approved by INQUA at its XVII Congress in Cairns, Australia, in 2007, the scale became officially known as *ESI 2007*.

Like many other intensity scales, ESI 2007 uses the basic structure of twelve degrees of seismic intensity and is designed for application during field surveys immediately after the seismic event. However, the definitions of intensity degrees in ESI 2007 are based on the observation of distribution and size of environmental effects produced by an earthquake. This approach makes ESI 2007 a unique diagnostic tool for the assessment of seismic intensity levels X to XII in sparsely populated and uninhibited areas where earthquake effects on people and built environment may not be easily observed. For intensity level IX or lower, the ESI 2007 scale is intended to be used as a supplement to other intensity scales.

a) the Definition of intensity degrees on the basis of Earthquake Environmental Effects;

b) the Guidelines, which aim at better clarifying i) the background of the scale and the scientific concepts that support the introduction of such a new macroseismic scale; ii) the procedure to use the scale alone or integrated with damage-based, traditional scales; iii) how the scale is organized; iv) the descriptions of diagnostic features required for intensity assessment, and the meaning of idioms, colors, and fonts.

Since 2007, the scale has been applied to modern, historical and paleoearthquakes in the frame of the INQUA TERPRO Commission on Paleoseismology and Active tectonics. It is now available in ten different languages.

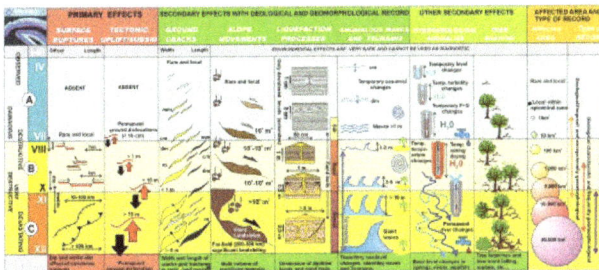

References

- Reitherman, Robert (2012). Earthquakes and Engineers: An International History. Reston, VA: ASCE Press. pp. 208–209. ISBN 9780784410714.

- Silver, Nate (2013). The signal and the noise : the art and science of prediction. London: Penguin. ISBN 9780141975658.

- Reitherman, Robert (2012). Earthquakes and Engineers: An International History. Reston, VA: ASCE Press. pp. 126–127. ISBN 9780784410714.

- "USGS Earthquake Magnitude Policy (implemented on January 18, 2002)". United States Geological Survey. January 30, 2014.

- Woo, Wang-chun (September 2012). "On Earthquake Magnitudes". Hong Kong Observatory. Retrieved 18 December 2013.

- "Anchorage, Alaska (AK) profile: population, maps, real estate, averages, homes, statistics, relocation, travel, jobs, hospitals, schools, crime, moving, houses, news". City-Data.com. Retrieved 2012-10-12.

Seismology: A Scientific Study

The study of seismic events, their causes, effects and prediction is known as seismology. This chapter provides comprehensive information on seismology, paleoseismology and seismometers and seismic waves. The content elaborately explores the fields and lists the types, methodology and terminology related to each of these topics.

Seismology

Seismology is the scientific study of earthquakes and the propagation of elastic waves through the Earth or through other planet-like bodies. The field also includes studies of earthquake environmental effects, such as tsunamis as well as diverse seismic sources such as volcanic, tectonic, oceanic, atmospheric, and artificial processes (such as explosions). A related field that uses geology to infer information regarding past earthquakes is paleoseismology. A recording of earth motion as a function of time is called a seismogram. A seismologist is a scientist who does research in seismology.

History

Scholarly interest in earthquakes can be traced back to antiquity. Early speculations on the natural causes of earthquakes were included in the writings of Thales of Miletus (c. 585 BCE), Anaximenes of Miletus (c. 550 BCE), Aristotle (c. 340 BCE) and Zhang Heng (132 CE).

In 132 CE, Zhang Heng of China's Han dynasty designed the first known seismoscope.

In 1664, Athanasius Kircher argued that earthquakes were caused by the movement of fire within a system of channels inside the Earth.

In 1703, Martin Lister (1638 to 1712) and Nicolas Lemery (1645 to 1715) proposed that earthquakes were caused by chemical explosions within the earth.

The Lisbon earthquake of 1755, coinciding with the general flowering of science in Europe, set in motion intensified scientific attempts to understand the behaviour and causation of earthquakes. The earliest responses include work by John Bevis (1757) and John Michell (1761). Michell determined that earthquakes originate within the Earth and were waves of movement caused by "shifting masses of rock miles below the surface."

From 1857, Robert Mallet laid the foundation of instrumental seismology and carried out seismological experiments using explosives. He is also responsible for coining the word "seismology".

In 1897, Emil Wiechert's theoretical calculations led him to conclude that the Earth's interior consists of a mantle of silicates, surrounding a core of iron.

In 1906 Richard Dixon Oldham identified the separate arrival of P-waves, S-waves and surface waves on seismograms and found the first clear evidence that the Earth has a central core.

In 1910, after studying the 1906 San Francisco earthquake, Harry Fielding Reid put forward the "elastic rebound theory" which remains the foundation for modern tectonic studies. The development of this theory depended on the considerable progress of earlier independent streams of work on the behaviour of elastic materials and in mathematics.

In 1926, Harold Jeffreys was the first to claim, based on his study of earthquake waves, that below the crust, the core of the Earth is liquid.

In 1937, Inge Lehmann determined that within the earth's liquid outer core there is a solid *inner* core.

By the 1960s, earth science had developed to the point where a comprehensive theory of the causation of seismic events had come together in the now well-established theory of plate tectonics.

Types of Seismic Wave

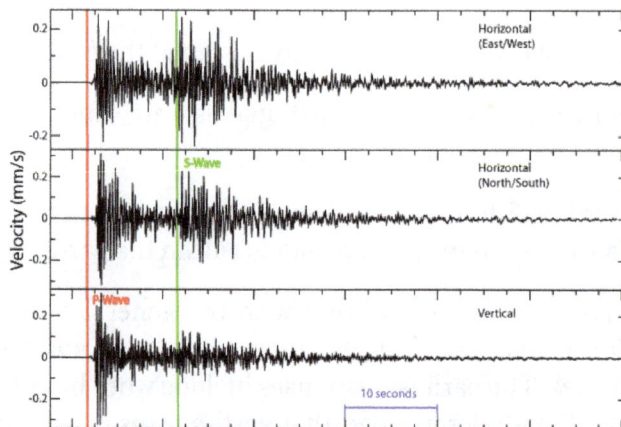

Seismogram records showing the three components of ground motion. The red line marks the first arrival of P-waves; the green line, the later arrival of S-waves.

Seismic waves are elastic waves that propagate in solid or fluid materials. They can be divided into *body waves* that travel through the interior of the materials; *surface waves* that travel along surfaces or interfaces between materials; and *normal modes*, a form of standing wave.

Body Waves

There are two types of body waves, Pressure waves or Primary waves (P-waves) and Shear or Secondary waves (S-waves). P-waves, are longitudinal waves that involve compression and expansion in the direction that the wave is moving. P-waves are the fastest waves in solids and are therefore the first waves to appear on a seismogram. S-waves are transverse waves that move perpendicular to the direction of propagation. S-waves are slower than P-waves. Therefore, they appear later than P-waves on a seismogram. Fluids cannot support perpendicular motion, so S-waves only travel in solids.

Surface Waves

The two main surface wave types are Rayleigh waves, which have some compressional motion, and Love waves, which do not. Rayleigh waves result from the interaction of vertically polarized P- and S-waves that satisfy the boundary conditions on the surface. Love waves can exist in the presence of a subsurface layer, and are only formed by horizontally polarized S-waves. Surface waves travel more slowly than P-waves and S-waves; however, because they are guided by the Earth's surface and their energy is thus trapped near the surface, they can be much stronger than body waves, and can be the largest signals on earthquake seismograms. Surface waves are strongly excited when their source is close to the surface, as in a shallow earthquake or a near surface explosion.

Normal Modes

Both body and surface waves are traveling waves; however, large earthquakes can also make the Earth "ring" like a bell. This ringing is a mixture of normal modes with discrete frequencies and periods of an hour or shorter. Motion caused by a large earthquake can be observed for up to a month after the event. The first observations of normal modes were made in the 1960s as the advent of higher fidelity instruments coincided with two of the largest earthquakes of the 20th century - the 1960 Valdivia earthquake and the 1964 Alaska earthquake. Since then, the normal modes of the Earth have given us some of the strongest constraints on the deep structure of the Earth.

Earthquakes

One of the first attempts at the scientific study of earthquakes followed the 1755 Lisbon earthquake. Other notable earthquakes that spurred major advancements in the science of seismology include the 1857 Basilicata earthquake, 1906 San Francisco earthquake, the 1964 Alaska earthquake, the 2004 Sumatra-Andaman earthquake, and the 2011 Great East Japan earthquake.

Controlled Seismic Sources

Seismic waves produced by explosions or vibrating controlled sources are one of the primary methods of underground exploration in geophysics (in addition to many different electromagnetic methods such as induced polarization and magnetotellurics). Controlled-source seismology has been used to map salt domes, anticlines and other geologic traps in petroleum-bearing rocks, faults, rock types, and long-buried giant meteor craters. For example, the Chicxulub Crater, which was caused by an impact that has been implicated in the extinction of the dinosaurs, was localized to Central America by analyzing ejecta in the Cretaceous–Paleogene boundary, and then physically proven to exist using seismic maps from oil exploration.

Detection of Seismic Waves

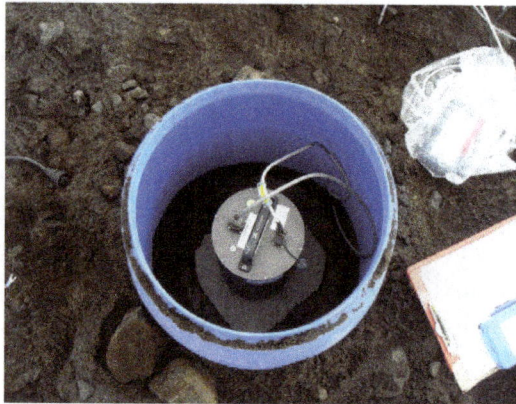

Installation for a temporary seismic station, north Iceland highland.

Seismometers are sensors that sense and record the motion of the Earth arising from elastic waves. Seismometers may be deployed at the Earth's surface, in shallow vaults, in boreholes, or underwater. A complete instrument package that records seismic signals is called a seismograph. Networks of seismographs continuously record ground motions around the world to facilitate the monitoring and analysis of global earthquakes and other sources of seismic activity. Rapid location of earthquakes makes tsunami warnings possible because seismic waves travel considerably faster than tsunami waves. Seismometers also record signals from non-earthquake sources ranging from explosions (nuclear and chemical), to local noise from wind or anthropogenic activities, to incessant signals generated at the ocean floor and coasts induced by ocean waves (the global microseism), to cryospheric events associated with large icebergs and glaciers. Above-ocean meteor strikes with energies as high as 4.2×10^{13} J (equivalent to that released by an explosion of ten kilotons of TNT) have been recorded by seismographs, as have a number of industrial accidents and terrorist bombs and events (a field of study referred to as forensic seismology). A major long-term motivation for the global seismographic monitoring has been for the detection and study of nuclear testing.

Mapping the Earth's Interior

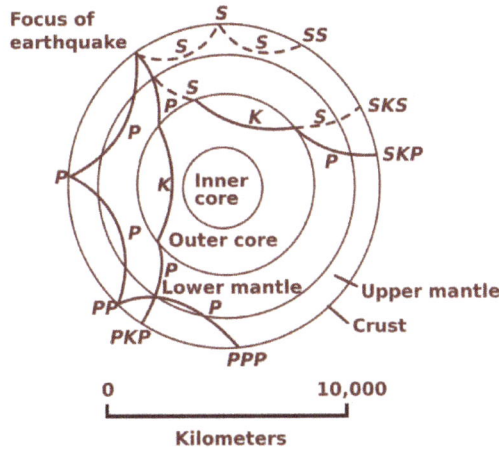

Seismic velocities and boundaries in the interior of the Earth sampled by seismic waves

Because seismic waves commonly propagate efficiently as they interact with the internal structure of the Earth, they provide high-resolution noninvasive methods for studying the planet's interior. One of the earliest important discoveries (suggested by Richard Dixon Oldham in 1906 and definitively shown by Harold Jeffreys in 1926) was that the outer core of the earth is liquid. Since S-waves do not pass through liquids, the liquid core causes a "shadow" on the side of the planet opposite of the earthquake where no direct S-waves are observed. In addition, P-waves travel much slower through the outer core than the mantle.

Processing readings from many seismometers using seismic tomography, seismologists have mapped the mantle of the earth to a resolution of several hundred kilometers. This has enabled scientists to identify convection cells and other large-scale features such as the large low-shear-velocity provinces near the core–mantle boundary.

Seismology and Society

Earthquake Prediction

Forecasting a probable timing, location, magnitude and other important features of a forthcoming seismic event is called earthquake prediction. Various attempts have been made by seismologists and others to create effective systems for precise earthquake predictions, including the VAN method. Most seismologists do not believe that a system to provide timely warnings for individual earthquakes has yet been developed, and many believe that such a system would be unlikely to give useful warning of impending seismic events. However, more general forecasts routinely predict seismic hazard. Such forecasts estimate the probability of an earthquake of a particular size affecting a particular location within a particular time-span, and they are routinely used in earthquake engineering.

Public controversy over earthquake prediction erupted after Italian authorities indicted six seismologists and one government official for manslaughter in connection with a magnitude 6.3 earthquake in L'Aquila, Italy on April 5, 2009. The indictment has been widely perceived as an indictment for failing to predict the earthquake and has drawn condemnation from the American Association for the Advancement of Science and the American Geophysical Union. The indictment claims that, at a special meeting in L'Aquila the week before the earthquake occurred, scientists and officials were more interested in pacifying the population than providing adequate information about earthquake risk and preparedness.

Engineering Seismology

Engineering seismology is the study and application of seismology for engineering purposes. It generally applied to the branch of seismology that deals with the assessment of the seismic hazard of a site or region for the purposes of earthquake engineering. It is, therefore, a link between earth science and civil engineering. There are two principal components of engineering seismology. Firstly, studying earthquake history (e.g. historical and instrumental catalogs of seismicity) and tectonics to assess the earthquakes that could occur in a region and their characteristics and frequency of occurrence. Secondly, studying strong ground motions generated by earthquakes to assess the expected shaking from future earthquakes with similar characteristics. These strong ground motions could either be observations from accelerometers or seismometers or those simulated by computers using various techniques.

Paleoseismology

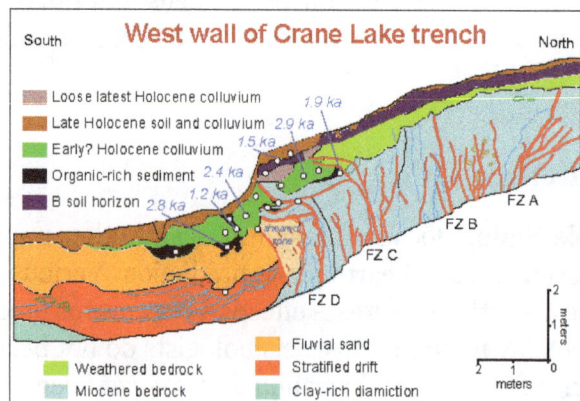

Sketch of trench wall

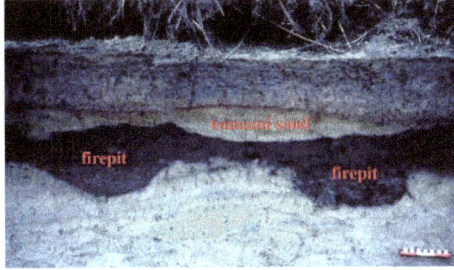

Effects of tsunami caused by an earthquake January, 26, 1700

Seismite formed by liquefaction of sediments during a Late Ordovician
earthquake (northern Kentucky, USA).

Paleoseismology looks at geologic sediments and rocks, for signs of ancient earthquakes. It is used to supplement seismic monitoring, for the calculation of seismic hazard. Paleoseismology is usually restricted to geologic regimes that have undergone continuous sediment creation for the last few thousand years, such as swamps, lakes, river beds and shorelines.

In this typical example, a trench is dug in an active sedimentation regime. Evidence of thrust faulting can be seen in the walls of the trench. It becomes a matter of deducting the relative age of each fault, by cross-cutting patterns. The faults can be dated in absolute terms, if there is dateable carbon, or human artifacts.

Many notable discoveries have been made using the techniques of paleoseismology. For example, there is a common misconception that having many smaller earthquakes can somehow 'relieve' a major fault such as the San Andreas, and reduce the chance of a major earthquake. It is now known (using paleoseismology) that nearly all the movement of the fault takes place with extremely large earthquakes. All of these seismic events (with a Moment Magnitude of over 8), leave some sort of trace in the sedimentation record.

Another famous example involves the Megathrust earthquakes of the Pacific Northwest. It was thought for some time that there was low seismic hazard in region because relatively few modern earthquakes are being recorded. There was a concept that the subduction zone was merely sliding in a benign manner.

All of these comforting notions were shattered by paleoseismology studies showing evidence of extremely large earthquakes, along with historical tsunami records.
In effect, the subduction zone under British Columbia, Washington, Oregon, and
far northern California, is perfectly normal, being extremely hazardous in the long
term, with the capability of generating coastal tsunamis of several hundred feet in
height at the coast. These are caused by the interface between the subducted sea
floor stressing the overlaying coastal soils in compression. Periodically a slip will
occur which causes the coastal portion to reduce in elevation and thrust toward
the west, leading to tsunamis in the central and eastern north Pacific ocean (with
several hours of warning) and a reflux of water toward the coastal shore, with little
time for residents to escape.

An Educational Excavation

Here on the Hayward Fault a pit has been dug for public educational purposes.
Click image for more information

Multiple image view from the platform.
The fault has been marked with cordage and various features labeled.

Image with enhanced annotation

Paleoseismic Trenching

Paleoseismic investigations are commonly performed through trenching studies in which a trench is dug and a geologist logs the geological attributes of the rock layers. Trenching studies are especially relevant to seismically active regions, such as many parts of California.

Seismometer

Japan Meteorological Agency Optical Electromagnetic Seismometer

Seismometers are instruments that measure motion of the ground, including those of seismic waves generated by earthquakes, volcanic eruptions, and other seismic sources. Records of seismic waves allow seismologists to map the interior of the Earth, and locate and measure the size of these different sources.

Seismograph is often used to mean *seismometer*, though it is more applicable to the older instruments in which the measuring and recording of ground motion were combined than to modern systems, in which these functions are separated. Both types provide a continuous record of ground motion; this distinguishes them from seismoscopes, which merely indicate that motion has occurred, perhaps with some simple measure of how large it was.

The concerning technical discipline is called seismometry, a branch of seismology.

Basic Principles

A simple seismometer that is sensitive to up-down motions of the earth can be understood by visualizing a weight hanging on a spring. The spring and weight are suspended from a frame that moves along with the earth surface. As the earth moves, the relative motion between the weight and the earth provides a measure of the vertical ground motion. If a recording system is installed, such as a rotating drum attached to the frame, and a pen attached to the mass, this relative motion between the weight and earth can be recorded to produce a history of ground motion, called a seismogram.

Any movement of the ground moves the frame. The mass tends not to move because of its inertia, and by measuring the movement between the frame and the mass, the motion of the ground can be determined.

Early seismometers used optical levers or mechanical linkages to amplify the small motions involved, recording on soot-covered paper or photographic paper. Modern instruments use electronics. In some systems, the mass is held nearly motionless relative to the frame by an electronic negative feedback loop. The motion of the mass relative to the frame is measured, and the feedback loop applies a magnetic or electrostatic force to keep the mass nearly motionless. The voltage needed to produce this force is the output of the seismometer, which is recorded digitally. In other systems the weight is allowed to move, and its motion produces a voltage in a coil attached to the mass and moving through the magnetic field of a magnet attached to the frame. This design is often used in the geophones used in seismic surveys for oil and gas.

Professional seismic observatories usually have instruments measuring three axes: north-south (y-axis), east-west (x-axis), and the vertical (z-axis). If only one axis is measured, this is usually the vertical because it is less noisy and gives better records of some seismic waves.

The foundation of a seismic station is critical. A professional station is sometimes mounted on bedrock. The best mountings may be in deep boreholes, which avoid thermal effects, ground noise and tilting from weather and tides. Other instruments are often mounted in insulated enclosures on small buried piers of unreinforced concrete. Reinforcing rods and aggregates would distort the pier as the temperature changes. A site is always surveyed for ground noise with a temporary installation before pouring the pier and laying conduit. Originally, European seismographs were placed in a particular area after a destructive earthquake. Today, they are spread to provide appropriate coverage (in the case of weak-motion seismology) or concentrated in high-risk regions (strong-motion seismology).

History

Ancient Era

In AD 132, Zhang Heng of China's Han dynasty invented the first seismoscope (by the definition above), which was called *Houfeng Didong Yi* (translated as, "instrument for measuring the seasonal winds and the movements of the Earth"). The description we have, from the History of the Later Han Dynasty, says that it was a large bronze vessel, about 2 meters in diameter; at eight points around the top were dragon's heads holding bronze balls. When there was an earthquake, one of the mouths would open and drop its ball into a bronze toad at the base, making a sound and supposedly showing the direction of the earthquake. On at least one occasion, probably at the time of a large earthquake in Gansu in AD 143, the seismoscope indicated an earthquake even though one was not felt. The available text says that inside the vessel was a central column that could move along eight tracks; this is thought to refer to a pendulum, though it is not known exactly how this was linked to a mechanism that would open only one dragon's mouth. The first ever earthquake recorded by this seismoscope was supposedly *somewhere in the east*. Days later, a rider from the east reported this earthquake.

Replica of Zhang Heng's seismoscope *Houfeng Didong Yi*

Modern Designs

The principle can be shown by an early special purpose seismometer. This consisted of a large stationary pendulum, with a stylus on the bottom. As the earth starts to move, the heavy mass of the pendulum has the inertia to stay still in the non-earth frame of reference. The result is that the stylus scratches a pattern corresponding with the Earth's movement. This type of strong motion seismometer recorded upon a smoked glass (glass with carbon soot). While not sensitive enough to detect distant earthquakes, this instrument could indicate the direction of the pressure waves and thus help find the epicenter of a local earthquake – such instruments were useful in the analysis of the 1906 San Francisco earthquake. Further re-analysis was performed in the 1980s using these early recordings, enabling a more precise determination of the initial fault break location in Marin county and its subsequent progression, mostly to the south.

Milne horizontal pendulum seismometer. One of the Important Cultural Properties of Japan.
Exhibit in the National Museum of Nature and Science, Tokyo, Japan.

After 1880, most seismometers were descended from those developed by the team of John Milne, James Alfred Ewing and Thomas Gray, who worked in Japan from 1880 to 1895. These seismometers used damped horizontal pendulums. After World War II, these were adapted into the widely used Press-Ewing seismometer.

Later, professional suites of instruments for the world-wide standard seismographic network had one set of instruments tuned to oscillate at fifteen seconds, and the other at ninety seconds, each set measuring in three directions. Amateurs or observatories with limited means tuned their smaller, less sensitive instruments to ten seconds. The basic damped horizontal pendulum seismometer swings like the gate of a fence. A heavy weight is mounted on the point of a long (from 10 cm to several meters) triangle, hinged at its vertical edge. As the ground moves, the weight stays unmoving, swinging the "gate" on the hinge.

The advantage of a horizontal pendulum is that it achieves very low frequencies of oscillation in a compact instrument. The "gate" is slightly tilted, so the weight tends to slowly return to a central position. The pendulum is adjusted (before the damping is installed) to oscillate once per three seconds, or once per thirty seconds. The general-purpose instruments of small stations or amateurs usually oscillate once per ten seconds. A pan of oil is placed under the arm, and a small sheet of metal mounted on the underside of the arm drags in the oil to damp oscillations. The level of oil, position on the arm, and angle and size of sheet is adjusted until the damping is "critical," that is, almost having oscillation. The hinge is very low friction, often torsion wires, so the only friction is the internal friction of the wire. Small seismographs with low proof masses are placed in a vacuum to reduce disturbances from air currents.

Zollner described torsionally suspended horizontal pendulums as early as 1869, but developed them for gravimetry rather than seismometry.

Early seismometers had an arrangement of levers on jeweled bearings, to scratch smoked glass or paper. Later, mirrors reflected a light beam to a direct-recording plate or roll of photographic paper. Briefly, some designs returned to mechanical movements to save money. In mid-twentieth-century systems, the light was reflected to a pair of differential electronic photosensors called a photomultiplier. The voltage generated in the photomultiplier was used to drive galvanometers which had a small mirror mounted on the axis. The moving reflected light beam would strike the surface of the turning drum, which was covered with photo-sensitive paper. The expense of developing photo sensitive paper caused many seismic observatories to switch to ink or thermal-sensitive paper.

Modern Instruments

CMG-40T triaxial broadband seismometer

Modern instruments use electronic sensors, amplifiers, and recording devices. Most are broadband covering a wide range of frequencies. Some seismometers can measure motions with frequencies from 500 Hz to 0.00118 Hz ($1/500 = 0.002$ seconds per cycle, to $1/0.00118 = 850$ seconds per cycle). The mechanical suspension for horizontal instruments remains the garden-gate described above. Vertical instruments use some kind of constant-force suspension, such as the LaCoste suspension. The LaCoste suspension uses a zero-length spring to provide a long period (high sensitivity). Some modern instruments use a "triaxial" design, in which three identical motion sensors are set at the same angle to the vertical but 120 degrees apart on the horizontal. Vertical and horizontal motions can be computed from the outputs of the three sensors.

Seismometers unavoidably introduce some distortion into the signals they measure, but professionally designed systems have carefully characterized frequency transforms.

Modern sensitivities come in three broad ranges: geophones, 50 to 750 V/m; local geologic seismographs, about 1,500 V/m; and teleseismographs, used for world survey, about 20,000 V/m. Instruments come in three main varieties: short period, long period and broadband. The short and long period measure velocity and are very sensitive, however they 'clip' the signal or go off-scale for ground motion that is strong enough to be felt by people. A 24-bit analog-to-digital conversion channel is commonplace. Practical devices are linear to roughly one part per million.

Delivered seismometers come with two styles of output: analog and digital. Analog seis-mographs require analog recording equipment, possibly including an analog-to-digital converter. The output of a digital seismograph can be simply input to a computer. It presents the data in a standard digital format (often "SE2" over Ethernet).

Teleseismometers

A low-frequency 3-direction ocean-bottom seismometer (cover removed). Two masses for x- and y-direction can be seen, the third one for z-direction is below. This model is a CMG-40TOBS, manufactured by Güralp Systems Ltd and is part of the Monterey Accelerated Research System.

The modern broadband seismograph can record a very broad range of frequencies. It consists of a small "proof mass", confined by electrical forces, driven by sophisticated electronics. As the earth moves, the electronics attempt to hold the mass steady through a feedback circuit. The amount of force necessary to achieve this is then recorded.

In most designs the electronics holds a mass motionless relative to the frame. This device is called a "force balance accelerometer". It measures acceleration instead of velocity of ground movement. Basically, the distance between the mass and some part of the frame is measured very precisely, by a linear variable differential transformer. Some instruments use a linear variable differential capacitor.

That measurement is then amplified by electronic amplifiers attached to parts of an electronic negative feedback loop. One of the amplified currents from the negative feedback loop drives a coil very like a loudspeaker, except that the coil is attached to the mass, and the magnet is mounted on the frame. The result is that the mass stays nearly motionless.

Most instruments measure directly the ground motion using the distance sensor. The voltage generated in a sense coil on the mass by the magnet directly measures the in-stantaneous velocity of the ground. The current to the drive coil provides a sensitive, accurate measurement of the force between the mass and frame, thus measuring di-rectly the ground's acceleration (using f=ma where f=force, m=mass, a=acceleration).

One of the continuing problems with sensitive vertical seismographs is the buoyancy of their masses. The uneven changes in pressure caused by wind blowing on an open window can easily change the density of the air in a room enough to cause a vertical seismograph to show spurious signals. Therefore, most professional seismographs are sealed in rigid gas-tight enclosures. For example, this is why a common Streckeisen model has a thick glass base that must be glued to its pier without bubbles in the glue.

It might seem logical to make the heavy magnet serve as a mass, but that subjects the seismograph to errors when the Earth's magnetic field moves. This is also why seismograph's moving parts are constructed from a material that interacts minimally with magnetic fields. A seismograph is also sensitive to changes in temperature so many instruments are constructed from low expansion materials such as nonmagnetic invar.

The hinges on a seismograph are usually patented, and by the time the patent has expired, the design has been improved. The most successful public domain designs use thin foil hinges in a clamp.

Another issue is that the transfer function of a seismograph must be accurately characterized, so that its frequency response is known. This is often the crucial difference between professional and amateur instruments. Most instruments are characterized on a variable frequency shaking table.

Strong-motion Seismometers

Another type of seismometer is a digital strong-motion seismometer, or accelerograph. The data from such an instrument is essential to understand how an earthquake affects manmade structures.

A strong-motion seismometer measures acceleration. This can be mathematically integrated later to give velocity and position. Strong-motion seismometers are not as sensitive to ground motions as teleseismic instruments but they stay on scale during the strongest seismic shaking.

Other Forms

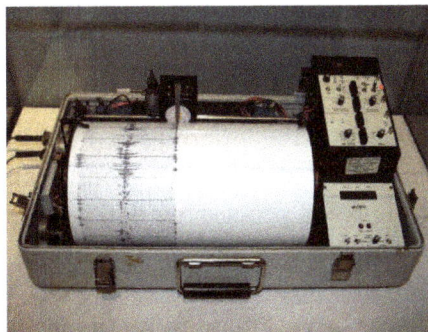

A Kinemetrics seismograph, formerly used by the United States Department of the Interior.

Accelerographs and geophones are often heavy cylindrical magnets with a spring-mounted coil inside. As case moves, the coil tends to stay stationary, so the magnetic field cuts the wires, inducing current in the output wires. They receive frequencies from several hundred hertz down to 1 Hz. Some have electronic damping, a low-budget way to get some of the performance of the closed-loop wide-band geologic seismographs.

Strain-beam accelerometers constructed as integrated circuits are too insensitive for geologic seismographs (2002), but are widely used in geophones.

Some other sensitive designs measure the current generated by the flow of a non-corrosive ionic fluid through an electret sponge or a conductive fluid through a magnetic field.

Interconnected Seismometers

Seismometers spaced in an array can also be used to precisely locate, in three dimensions, the source of an earthquake, using the time it takes for seismic waves to propagate away from the hypocenter, the initiating point of fault rupture. Interconnected seismometers are also used to detect underground nuclear test explosions, as well as for Earthquake early warning systems. These seismometer are often used as part of a large scale governmental or scientific project, but some organizations such as the Quake-Catcher Network, can use residential size detectors built into computers to detect earthquakes as well.

In reflection seismology, an array of seismometers image sub-surface features. The data are reduced to images using algorithms similar to tomography. The data reduction methods resemble those of computer-aided tomographic medical imaging X-ray machines (CAT-scans), or imaging sonars.

A world-wide array of seismometers can actually image the interior of the Earth in wave-speed and transmissivity. This type of system uses events such as earthquakes, impact events or nuclear explosions as wave sources. The first efforts at this method used manual data reduction from paper seismograph charts. Modern digital seismograph records are better adapted to direct computer use. With inexpensive seismometer designs and internet access, amateurs and small institutions have even formed a "public seismograph network."

Seismographic systems used for petroleum or other mineral exploration historically used an explosive and a wireline of geophones unrolled behind a truck. Now most short-range systems use "thumpers" that hit the ground, and some small commercial systems have such good digital signal processing that a few sledgehammer strikes provide enough signal for short-distance refractive surveys. Exotic cross or two-dimensional arrays of geophones are sometimes used to perform three-dimensional reflective imaging of subsurface features. Basic linear refractive geomapping software (once a black art) is available off-the-shelf, running on laptop computers, using strings as small as three geophones. Some systems now come in an 18" (0.5 m) plastic field case with a computer, display and printer in the cover.

Small seismic imaging systems are now sufficiently inexpensive to be used by civil engineers to survey foundation sites, locate bedrock, and find subsurface water.

Recording

Viewing of a Develocorder film

Matsushiro Seismological Observatory

Today, the most common recorder is a computer with an analog-to-digital converter, a disk drive and an internet connection; for amateurs, a PC with a sound card and associated software is adequate. Most systems record continuously, but some record only when a signal is detected, as shown by a short-term increase in the variation of the signal, compared to its long-term average (which can vary slowly because of changes in seismic noise).

Prior to the availability of digital processing of seismic data in the late 1970s, the records were done in a few different forms on different types of media. A "Helicorder" drum was a device used to record data into photographic paper or in the form of paper and ink. A "Develocorder" was a machine that record data from up to 20 channels into a 16-mm film. The recorded film can be viewed by a machine. The reading and measuring from these types of media can be done by hand. After the digital processing has been used, the archives of the seismic data were recorded in magnetic tapes. Due to the deterioration of older magnetic tape medias, large number of waveforms from the archives are not recoverable.

Seismic Wave

Seismic waves are waves of energy that travel through the Earth's layers, and are a result of earthquakes, volcanic eruptions, magma movement, large landslides and large man-made explosions that give out low-frequency acoustic energy. Many other natural and anthropogenic sources create low-amplitude waves commonly referred to as ambient vibrations. Seismic waves are studied by geophysicists called seismologists. Seismic wave fields are recorded by a seismometer, hydrophone (in water), or accelerometer.

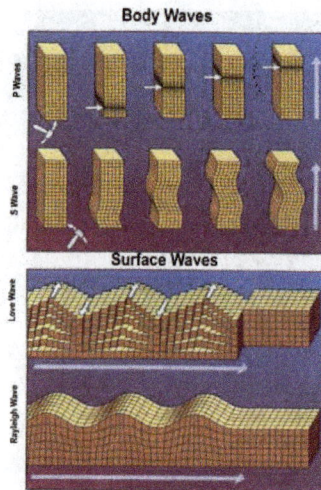

Body waves and surface waves

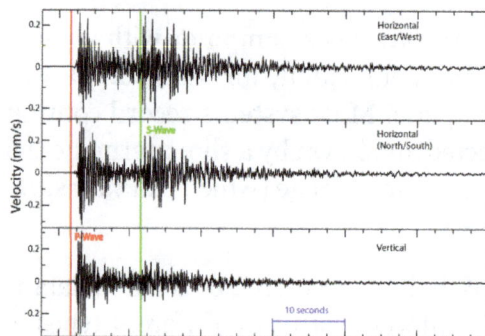

p-wave and s-wave from seismograph

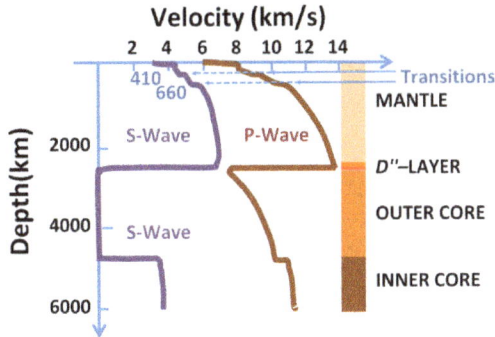

Velocity of seismic waves in the Earth versus depth. The negligible *S*-wave velocity in the outer core occurs because it is liquid, while in the solid inner core the *S*-wave velocity is non-zero.

The propagation velocity of the waves depends on density and elasticity of the medium. Velocity tends to increase with depth and ranges from approximately 2 to 8 km/s in the Earth's crust, up to 13 km/s in the deep mantle.

Earthquakes create distinct types of waves with different velocities; when reaching seismic observatories, their different travel times help scientists to locate the source of the hypocenter. In geophysics the refraction or reflection of seismic waves is used for research into the structure of the Earth's interior, and man-made vibrations are often generated to investigate shallow, subsurface structures.

Types

Among the many types of seismic waves, one can make a broad distinction between *body waves*, which travel through the Earth, and *surface waves*, which travel at the Earth's surface.

Other modes of wave propagation exist than those described in this article; though of comparatively minor importance for earth-borne waves, they are important in the case of asteroseismology.

- Body waves travel through the interior of the Earth.

- Surface waves travel across the surface. Surface waves decay more slowly with distance than do body waves, which travel in three dimensions.

- Particle motion of surface waves is larger than that of body waves, so surface waves tend to cause more damage.

Body Waves

Body waves travel through the interior of the Earth along paths controlled by the material properties in terms of density and modulus (stiffness). The density and modulus, in turn, vary according to temperature, composition, and material phase. This effect resembles the refraction of light waves. Two types of particle motion result in two types of body waves: *Primary* and *Secondary* waves.

Primary Waves

Primary waves (P-waves) are compressional waves that are longitudinal in nature. P waves are pressure waves that travel faster than other waves through the earth to arrive at seismograph stations firstly, hence the name "Primary". These waves can travel through any type of material, including fluids, and can travel at nearly twice the speed of S waves. In air, they take the form of sound waves, hence they travel at the speed of sound. Typical speeds are 330 m/s in air, 1450 m/s in water and about 5000 m/s in granite.

Secondary Waves

Secondary waves (S-waves) are shear waves that are transverse in nature. Following an earthquake event, S-waves arrive at seismograph stations after the faster-moving P-waves and displace the ground perpendicular to the direction of propagation. Depending on the propagational direction, the wave can take on different surface characteristics; for example, in the case of horizontally polarized S waves, the ground moves alternately to one side and then the other. S-waves can travel only through solids, as fluids (liquids and gases) do not support shear stresses. S-waves are slower than P-waves, and speeds are typically around 60% of that of P-waves in any given material.

Surface Waves

Seismic surface waves travel along the Earth's surface. They can be classified as a form of mechanical surface waves. They are called surface waves, as they diminish as they get further from the surface. They travel more slowly than seismic body waves (P and S). In large earthquakes, surface waves can have an amplitude of several centimeters.

Rayleigh Waves

Rayleigh waves, also called ground roll, are surface waves that travel as ripples with motions that are similar to those of waves on the surface of water (note, however, that the associated particle motion at shallow depths is retrograde, and that the restoring force in Rayleigh and in other seismic waves is elastic, not gravitational as for water waves). The existence of these waves was predicted by John William Strutt, Lord Rayleigh, in 1885. They are slower than body waves, roughly 90% of the velocity of S waves for typical homogeneous elastic media. In the layered medium (like the crust and upper mantle) the velocity of the Rayleigh waves depends on their frequency and wavelength.

Love Waves

Love waves are horizontally polarized shear waves (SH waves), existing only in the presence of a semi-infinite medium overlain by an upper layer of finite thickness. They are named after A.E.H. Love, a British mathematician who created a mathematical model of the waves in 1911. They usually travel slightly faster than Rayleigh waves, about 90% of the S wave velocity, and have the largest amplitude.

Stoneley Waves

A Stoneley wave is a type of boundary wave (or interface wave) that propagates along a solid-fluid boundary or, under specific conditions, also along a solid-solid boundary. Amplitudes of Stoneley waves have their maximum values at the boundary between the two contacting media and decay exponentially towards the depth of each of them. These waves can be generated along the walls of a fluid-filled borehole, being an important source of coherent noise in VSPs and making up the low frequency component of the source in sonic logging. The equation for Stoneley waves was first given by Dr. Robert Stoneley (1894 - 1976), Emeritus Professor of Seismology, Cambridge.

Free Oscillations of the Earth

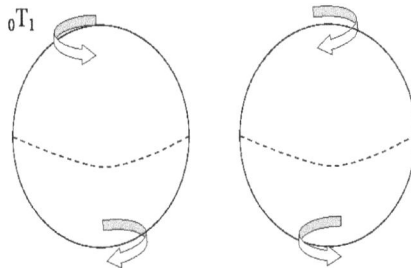

The sense of motion for toroidal $_0T_1$ oscillation for two moments of time.

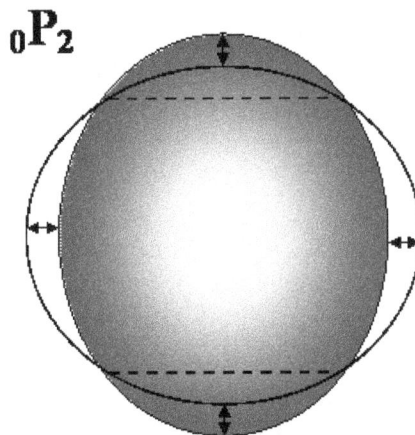

The scheme of motion for spheroidal $_0S_2$ oscillation. Dashed lines give nodal (zero) lines. Arrows give the sense of motion.

Free oscillations of the Earth are standing waves, the result of interference between two surface waves traveling in opposite directions. Interference of Rayleigh waves results in *spheroidal oscillation S* while interference of Love waves gives *toroidal oscillation T*. The modes of oscillations are specified by three numbers, e.g., $_nS_l^m$, where l is the an-gular order number (or *spherical harmonic degree*). The number m is the azimuthal order number. It may take on $2l+1$ values from $-l$ to $+l$. The number n is the *radial order number*. It means the wave with n zero crossings in radius. For spherically symmetric Earth the period for given n and l does not depend on m.

Some examples of spheroidal oscillations are the "breathing" mode $_0S_0$, which in-volves an expansion and contraction of the whole Earth, and has a period of about 20 minutes; and the "rugby" mode $_0S_2$, which involves expansions along two alter-nating directions, and has a period of about 54 minutes. The mode $_0S_1$ does not exist because it would require a change in the center of gravity, which would require an external force.

Of the fundamental toroidal modes, $_0T_1$ represents changes in Earth's rotation rate; although this occurs, it is much too slow to be useful in seismology. The mode $_0T_2$ de-scribes a twisting of the northern and southern hemispheres relative to each other; it has a period of about 44 minutes.

The first observations of free oscillations of the Earth were done during the great 1960 earthquake in Chile. Presently periods of thousands modes are known. These data are used for determining some large scale structures of the Earth interior.

P and S Waves in Earth's Mantle and Core

When an earthquake occurs, seismographs near the epicenter are able to record both P and S waves, but those at a greater distance no longer detect the high frequencies of the first S wave. Since shear waves cannot pass through liquids, this phenomenon was original evidence for the now well-established observation that the Earth has a liquid outer core, as demonstrated by Richard Dixon Oldham. This kind of observation has also been used to argue, by seismic testing, that the Moon has a solid core, although recent geodetic studies suggest the core is still molten.

Notation

The path that a wave takes between the focus and the observation point is often drawn as a ray diagram. An example of this is shown in a figure above. When reflections are taken into account there are an infinite number of paths that a wave can take. Each path is denoted by a set of letters that describe the trajectory and phase through the Earth. In general an upper case denotes a transmitted wave and a lower case denotes a reflected wave. The two exceptions to this seem to be "g" and "n".

c	the wave reflects off the outer core
d	a wave that has been reflected off a discontinuity at depth d
g	a wave that only travels through the crust
i	a wave that reflects off the inner core
I	a P-wave in the inner core
h	a reflection off a discontinuity in the inner core
J	an S wave in the inner core
K	a P-wave in the outer core
L	a Love wave sometimes called LT-Wave (Both caps, while an Lt is different)
n	a wave that travels along the boundary between the crust and mantle
P	a P wave in the mantle
p	a P wave ascending to the surface from the focus
R	a Rayleigh wave
S	an S wave in the mantle
s	an S wave ascending to the surface from the focus
w	the wave reflects off the bottom of the ocean
	No letter is used when the wave reflects off of the surfaces

For example:

- ScP is a wave that begins traveling towards the center of the Earth as an S wave. Upon reaching the outer core the wave reflects as a P wave.

- sPKIKP is a wave path that begins traveling towards the surface as an S-wave. At the surface it reflects as a P-wave. The P-wave then travels through the outer core, the inner core, the outer core, and the mantle.

Usefulness of P and S Waves in Locating an Event

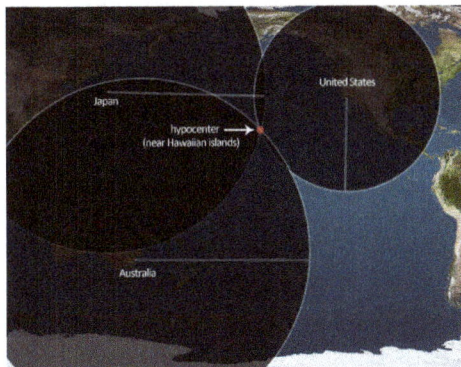

The Hypocenter/Epicenter of an earthquake is calculated by using the seismic data of that earthquake from at least three different locations.

In the case of local or nearby earthquakes, the difference in the arrival times of the P and S waves can be used to determine the distance to the event. In the case of earthquakes that have occurred at global distances, three or more geographically diverse observing stations (using a common clock) recording P-wave arrivals permits the computation of a unique time and location on the planet for the event. Typically, dozens or even hundreds of P-wave arrivals are used to calculate hypocenters. The misfit generated by a hypocenter calculation is known as "the residual". Residuals of 0.5 second or less are typical for distant events, residuals of 0.1-0.2 s typical for local events, meaning most reported P arrivals fit the computed hypocenter that well. Typically a location program will start by assuming the event occurred at a depth of about 33 km; then it minimizes the residual by adjusting depth. Most events occur at depths shallower than about 40 km, but some occur as deep as 700 km.

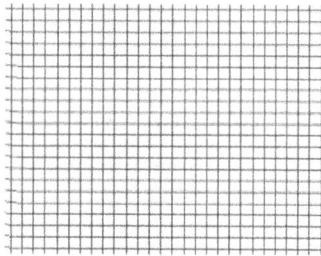

P- and S-waves sharing with the propagation

A quick way to determine the distance from a location to the origin of a seismic wave less than 200 km away is to take the difference in arrival time of the P wave and the S wave in seconds and multiply by 8 kilometers per second. Modern seismic arrays use more complicated earthquake location techniques.

At teleseismic distances, the first arriving P waves have necessarily travelled deep into the mantle, and perhaps have even refracted into the outer core of the planet, before travelling back up to the Earth's surface where the seismographic stations are located. The waves travel more quickly than if they had traveled in a straight line from the earthquake. This is due to the appreciably increased velocities within the planet, and is termed Huygens' Principle. Density in the planet increases with depth, which would slow the waves, but the modulus of the rock increases much more, so deeper means faster. Therefore, a longer route can take a shorter time.

The travel time must be calculated very accurately in order to compute a precise hypocenter. Since P waves move at many kilometers per second, being off on travel-time calculation by even a half second can mean an error of many kilometers in terms of distance. In practice, P arrivals from many stations are used and the errors cancel out, so the computed epicenter is likely to be quite accurate, on the order of 10–50 km or so around the world. Dense arrays of nearby sensors such as those that exist in California can provide accuracy of roughly a kilometer, and much greater accuracy is possible when timing is measured directly by cross-correlation of seismogram waveforms.

References

- Shearer, Peter M. (2009). Introduction to Seismology (Second ed.). Cambridge University Press. ISBN 978-0-521-70842-5.

- Stein, Seth; Wysession, Michael (2002). An Introduction to Seismology, Earthquakes and Earth Structure. Wiley-Blackwell. ISBN 978-0-86542-078-6.

- Ben-Menahem, A. (2009). Historical Encyclopedia of Natural and Mathematical Sciences , Volume 1. Springer. p. 2657. ISBN 9783540688310. Retrieved 28 August 2012.

- Reitherman, Robert (2012). Earthquakes and Engineers: an International History. Reston, VA: ASCE Press. pp. 122–125. ISBN 9780784410714.

  o Peter M. Shearer (2009). Introduction to Seismology. Cambridge University Press. ISBN 978-0-521-88210-1.

  o Seth Stein; Michael Wysession (1 April 2009). An Introduction to Seismology, Earthquakes, and Earth Structure. John Wiley & Sons. ISBN 978-14443-1131-0.

  o Sheriff, R. E., Geldart, L. P. (1995). Exploration Seismology (2nd ed.). Cambridge University Press. p. 52. ISBN 0-521-46826-4.

  o William H.K. Lee; Paul Jennings; Carl Kisslinger; Hiroo Kanamori (27 September 2002). International Handbook of Earthquake & Engineering Seismology. Academic Press. pp. 283–. ISBN 978-0-08-048922-3. Retrieved 29 April 2013.

  o Hutton, Kate; Yu, Ellen. "NEWS FLASH!! SCSN Earthquake Catalog Completed!!" (PDF). Seismological Laboratory, Caltech. Retrieved 4 July 2014.

5

Methods to Mitigate Seismic Vibrations

Seismic activity can cause an insurmountable amount of damage to human lives, cities and the environment as well. To protect lives and man-made structures has been the main focus of earthquake engineering. Three major techniques to achieve this have been discussed in this chapter- vibration control, vibration isolation and seismic retrofit. These techniques are applied by various devices that are fitted into buildings to mitigate the damage of earthquakes.

Vibration Control

In earthquake engineering, vibration control is a set of technical means aimed to mitigate seismic impacts in building and non-building structures.

All seismic vibration control devices may be classified as *passive, active* or *hybrid* where:

Base isolator being tested at the UCSD Caltrans-SRMD facility

- *passive control devices* have no feedback capability between them, structural elements and the ground;

- *active control devices* incorporate real-time recording instrumentation on the ground integrated with earthquake input processing equipment and actuators within the structure;

- *hybrid control devices* have combined features of active and passive control systems.

When ground seismic waves reach up and start to penetrate a base of a building, their energy flow density, due to reflections, reduces dramatically: usually, up to 90%. However, the remaining portions of the incident waves during a major earthquake still bear a huge devastating potential.

After the seismic waves enter a superstructure, there is a number of ways to control them in order to sooth their damaging effect and improve the building's seismic performance, for instance:

- to dissipate the wave energy inside a superstructure with properly engineered dampers;

- to disperse the wave energy between a wider range of frequencies;

- to absorb the resonant portions of the whole wave frequencies band with the help of so-called *mass dampers* .

Base-isolated San Francisco City Hall after seismic retrofit

Devices of the last kind, abbreviated correspondingly as TMD for the tuned (*passive*), as AMD for the *active*, and as HMD for the *hybrid mass dampers*, have been studied and installed in high-rise buildings, predominantly in Japan, for a quarter of a century .

However, there is quite another approach: partial suppression of the seismic energy flow into the superstructure known as *seismic* or *base isolation* which has been implemented in a number of historical buildings all over the world and remains in the focus of earthquake engineering research for years.

For this, some pads are inserted into all major load-carrying elements in the base of the building which should substantially decouple a superstructure from its substructure resting on a shaking ground. It also requires creating a rigidity diaphragm and a moat around the building, as well as making provisions against overturning and P-delta effect.

In refineries or plants snubbers are often used for vibration control. Snubbers come in two different variations: hydraulic snubber and a mechanical snubber.

- Hydraulic snubbers are used on piping systems when restrained thermal movement is allowed.

- Mechanical snubbers operate on the standards of restricting acceleration of any pipe movements to a threshold of 0.2 g's, which is the maximum acceleration that the snubber will permit the piping to see.

Vibration Isolation

Vibration isolation is the process of isolating an object, such as a piece of equipment, from the source of vibrations.

Vibration is undesirable in many domains, primarily engineered systems and habitable spaces, and methods have been developed to prevent the transfer of vibration to such systems. Vibrations propagate via mechanical waves and certain mechanical linkages conduct vibrations more efficiently than others. Passive vibration isolation makes use of materials and mechanical linkages that absorb and damp these mechanical waves. Active vibration isolation involves sensors and actuators that produce destructive interference that cancels-out incoming vibration.

Passive Isolation

"Passive vibration isolation" refers to vibration isolation or mitigation of vibrations by passive techniques such as rubber pads or mechanical springs, as opposed to "active vibration isolation" or "electronic force cancellation" employing electric power, sensors, actuators, and control systems.

Passive vibration isolation is a vast subject, since there are many types of passive vibration isolators used for many different applications. A few of these applications are for industrial equipment such as pumps, motors, HVAC systems, or washing machines; isolation of civil engineering structures from earthquakes (base isolation), sensitive laboratory equipment, valuable statuary, and high-end audio.

A basic understanding of how passive isolation works, the more common types of passive isolators, and the main factors that influence the selection of passive isolators:

Common Passive Isolation Systems

Pneumatic or Air Isolators

These are bladders or canisters of compressed air. A source of compressed air is required to maintain them. Air springs are rubber bladders which provide damping as well as isolation and are used in large trucks. Some pneumatic isolators can attain low resonant frequencies and are used for isolating large industrial equipment. Air tables consist of a working surface or optical surface mounted on air legs. These tables provide enough isolation for laboratory instrument under some conditions. Air systems may leak under vacuum conditions. The air container can interfere with isolation of low-amplitude vibration.

Mechanical Springs and Spring-dampers

These are heavy-duty isolators used for building systems and industry. Sometimes they serve as mounts for a concrete block, which provides further isolation.

Pads or sheets of flexible materials such as elastomers, rubber, cork, dense foam and laminate materials.

Elastomer pads, dense closed cell foams and laminate materials are often used under heavy machinery, under common household items, in vehicles and even under higher performing audio systems.

Molded and Bonded Rubber and Elastomeric Isolators and Mounts

These are often used as machinery mounts or in vehicles. They absorb shock and attenuate some vibration.

Negative-stiffness Isolators

Negative-stiffness isolators are less common than other types and have generally been developed for high-level research applications such as gravity wave detection. Lee, Goverdovskiy, and Temnikov (2007) proposed a negative-stiffness system for isolating vehicle seats.

The focus on negative 'stiffness isolators has been on developing systems with very low resonant frequencies (below 1 Hz), so that low frequencies can be adequately isolated, which is critical for sensitive instrumentation. All higher frequencies are also isolated. Negative stiffness systems can be made with low stiction, so that they are effective in isolating low-amplitude vibrations.

Negative-stiffness mechanisms are purely mechanical and typically involve the configuration and loading of components such as beams or inverted pendulums. Greater loading of the negative-stiffness mechanism, within the range of its operability, decreases the natural frequency.

Wire Rope Isolators

These isolators are durable and can withstand extreme environments. They are often used in military applications.

Base isolators for seismic isolation of buildings, bridges, etc.

> Base isolators made of layers of neoprene and steel with a low horizontal stiffness are used to lower the natural frequency of the building. Some other base isolators are designed to slide, preventing the transfer of energy from the ground to the building.

Tuned Mass Dampers

> Tuned mass dampers reduce the effects of harmonic vibration in buildings or other structures. A relatively small mass is attached in such a way that it can dampen out a very narrow band of vibration of the structure.

Do it Yourself Isolators

> In less sophisticated solutions, bungee cords can be used as a cheap isolation system which may be effective enough for some applications. The item to be isolated is suspended from the bungee cords. This is difficult to implement without a danger of the isolated item falling. Tennis balls cut in half have been used under washing machines and other items with some success.

How Passive Isolation Works

A passive isolation system, such as a shock mount, in general contains mass, spring, and damping elements and moves as a harmonic oscillator. The mass and spring stiffness dictate a natural frequency of the system. Damping causes energy dissipation and has a secondary effect on natural frequency.

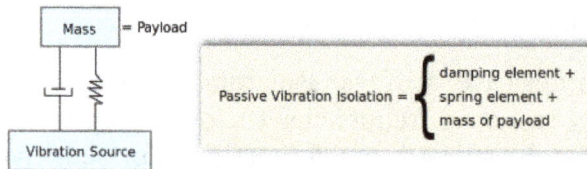

Passive Vibration Isolation

Every object on a flexible support has a fundamental natural frequency. When vibration is applied, energy is transferred most efficiently at the natural frequency, somewhat efficiently below the natural frequency, and with increasing inefficiency (decreasing efficiency) above the natural frequency. This can be seen in the transmissibility curve, which is a plot of transmissibility vs. frequency.

Here is an example of a transmissibility curve. Transmissibility is the ratio of vibration of the isolated surface to that of the source. Vibrations are never completely eliminated, but they can be greatly reduced. The curve below shows the typical performance of a passive, negative-stiffness isolation system with a natural frequency of 0.5 Hz. The general shape of the curve is typical for passive systems. Below the natural frequency, transmissibility hovers near 1. A value of 1 means that vibration is going through the system without being amplified or reduced. At the resonant frequency, energy is transmitted efficiently, and the incoming vibration is amplified. Damping in the system limits the level of amplification. Above the resonant frequency, little energy can be transmitted, and the curve rolls off to a low value. A passive isolator can be seen as a mechanical low-pass filter for vibrations.

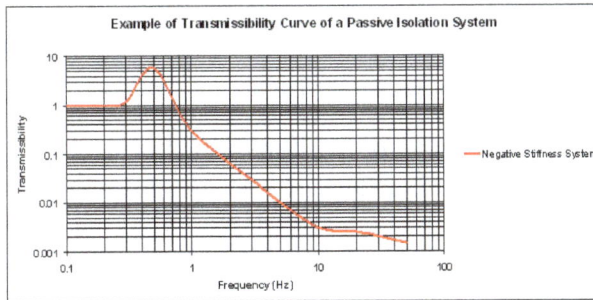

negative-stiffness transmissibility

In general, for any given frequency above the natural frequency, an isolator with a lower natural frequency will show greater isolation than one with a higher natural frequency. The best isolation system for a given situation depends on the frequency, direction, and magnitude of vibrations present and the desired level of attenuation of those frequencies.

All mechanical systems in the real world contain some amount of damping. Damping dissipates energy in the system, which reduces the vibration level which is transmitted at the natural frequency. The fluid in automotive shock absorbers is a kind of damper, as is the inherent damping in elastomeric (rubber) engine mounts.

Damping is used in passive isolators to reduce the amount of amplification at the natural frequency. However, increasing damping tends to reduce isolation at the higher frequencies. As damping is increased, transmissibility roll-off decreases. This can be seen in the chart below.

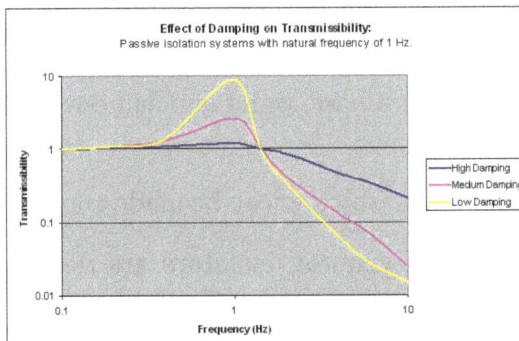

Damping effect on transmissibility

Passive isolation operates in both directions, isolating the payload from vibrations originating in the support, and also isolating the support from vibrations originating in the payload. Large machines such as washers, pumps, and generators, which would cause vibrations in the building or room, are often isolated from the floor. However, there are a multitude of sources of vibration in buildings, and it is often not possible to isolate each source. In many cases, it is most efficient to isolate each sensitive instrument from the floor. Sometimes it is necessary to implement both approaches.

Factors influencing the selection of passive vibration isolators

1. Characteristics of item to be isolated

    o   Size: The dimensions of the item to be isolated help determine the type of isolation which is available and appropriate. Small objects may use only one isolator, while larger items might use a multiple-isolator system.

    o   Weight: The weight of the object to be isolated is an important factor in choosing the correct passive isolation product. Individual passive isolators are designed to be used with a specific range of loading.

    o   Movement: Machines or instruments with moving parts may affect isolation systems. It is important to know the mass, speed, and distance traveled of the moving parts.

2. Operating Environment

    o   Industrial: This generally entails strong vibrations over a wide band of frequencies and some amount of dust.

    o   Laboratory: Labs are sometimes troubled by specific building vibrations from adjacent machinery, foot traffic, or HVAC airflow.

    o   Indoor or outdoor: Isolators are generally designed for one environment or the other.

    o   Corrosive/non-corrosive: Some indoor environments may present a corrosive danger to isolator components due to the presence of corrosive chemicals. Outdoors, water and salt environments need to be considered.

    o   Clean room: Some isolators can be made appropriate for clean room.

    o   Temperature: In general, isolators are designed to be used in the range of temperatures normal for human environments. If a larger range of temperatures is required, the isolator design may need to be modified.

    o   Vacuum: Some isolators can be used in a vacuum environment. Air isolators may have leakage problems. Vacuum requirements typically include some level of clean room requirement and may also have a large temperature range.

    o   Magnetism: Some experimentation which requires vibration isolation also requires a low-magnetism environment. Some isolators can be designed with low-magnetism components.

    o   Acoustic noise: Some instruments are sensitive to acoustic vibration. In addition, some isolation systems can be excited by acoustic noise. It may be necessary to use an acoustic shield. Air compressors can create problematic acoustic noise, heat, and airflow.

    o   Static or dynamic loads: This distinction is quite important as isolators are designed for a certain type and level of loading.

Static loading

is basically the weight of the isolated object with low-amplitude vibration input. This is the environment of apparently stationary objects such as buildings (under normal conditions) or laboratory instruments.

Dynamic loading

involves accelerations and larger amplitude shock and vibration. This environment is present in vehicles, heavy machinery, and structures with significant movement.

3. Cost:

    o   Cost of providing isolation: Costs include the isolation system itself, whether it is a standard or custom product; a compressed air source if required; shipping from manufacturer to destination; installation; maintenance; and an initial vibration site survey to determine the need for isolation.

    o   Relative costs of different isolation systems: Inexpensive shock mounts may need to be replaced due to dynamic loading cycles. A higher level of isolation which is effective at lower vibration frequencies and magnitudes generally costs more. Prices can range from a few dollars for bungee cords to millions of dollars for some space applications.

4. Adjustment: Some isolation systems require manual adjustment to compensate for changes in weight load, weight distribution, temperature, and air pressure, whereas other systems are designed to automatically compensate for some or all of these factors.

5. Maintenance: Some isolation systems are quite durable and require little or no maintenance. Others may require periodic replacement due to mechanical fatigue of parts or aging of materials.

6. Size Constraints: The isolation system may have to fit in a restricted space in a laboratory or vacuum chamber, or within a machine housing.

7. Nature of vibrations to be isolated or mitigated

    o Frequencies: If possible, it is important to know the frequencies of ambient vibrations. This can be determined with a site survey or accelerometer data processed through FFT analysis.

    o Amplitudes: The amplitudes of the vibration frequencies present can be compared with required levels to determine whether isolation is needed. In addition, isolators are designed for ranges of vibration amplitudes. Some isolators are not effective for very small amplitudes.

    o Direction: Knowing whether vibrations are horizontal or vertical can help to target isolation where it is needed and save money.

8. Vibration specifications of item to be isolated: Many instruments or machines have manufacturer-specified levels of vibration for the operating environment. The manufacturer may not guarantee the proper operation of the instrument if vibration exceeds the spec.

Comparison of Passive Isolators

Type of Passive Isolation	Applications	Typical Natural Frequency
Air Isolators	Large industrial equipment, some optics and instruments	1.5 - 3 Hz, large systems customized to 0.5 Hz
Springs or spring dampers	Heavy loads, pumps, compressors	3 - 9 Hz
Elastomer or cork pads	Large high-load applications where isolation of medium to high frequency noise and vibration is required	3 - 40 Hz, depending on size of pad and load
Molded or bonded elastomer mounts	Machinery, instruments, vehicles, aviation	10 - 20+ Hz
Negative stiffness isolators	Electron microscopes, sensitive instruments, optics and laser systems, cryogenic systems	0.17 - 2.5 Hz
Wire rope isolators	Machinery, instruments, vehicles, aviation	10 - 40+ Hz
Bungee cord isolators	Laboratory, home, etc.	Depends on type of cord and the mass they support
Base isolators	Buildings and large structures	Low, seismic frequencies
Tuned Mass Dampers	Buildings, large structures, aerospace	Any, but usually used at low frequencies

Negative-stiffness Vibration Isolator

Negative-Stiffness-Mechanism (NSM) vibration isolation systems offer a unique passive approach for achieving low vibration environments and isolation against sub-Hertz vibrations. "Snap-through" or "over-center" NSM devices are used to reduce the stiffness of elastic suspensions and create compact six-degree-of-freedom systems with low natural frequencies. Practical systems with vertical and horizontal natural frequencies as low as 0.2 to 0.5 Hz are possible. Electro-mechanical auto-adjust mechanisms compensate for varying weight loads and provide automatic leveling in multiple-isolator systems, similar to the function of leveling valves in pneumatic systems. All-metal systems can be configured which are compatible with high vacuums and other adverse environments such as high temperatures.

These isolation systems enable vibration-sensitive instruments such as scanning probe microscopes, micro-hardness testers and scanning electron microscopes to operate in severe vibration environments sometimes encountered, for example, on upper floors of buildings and in clean rooms. Such operation would not be practical with pneumatic isolation systems. Similarly, they enable vibration-sensitive instruments to produce better images and data than those achievable with pneumatic isolators.

The theory of operation of NSM vibration isolation systems is summarized, some typical systems and applications are described, and data on measured performance is presented. The theory of NSM isolation systems is explained in References 1 and 2. It is summarized briefly for convenience.

Vertical-motion Isolation

A vertical-motion isolator is shown . It uses a conventional spring connected to an NSM consisting of two bars hinged at the center, supported at their outer ends on pivots, and loaded in compression by forces P. The spring is compressed by weight W to the operating position of the isolator, as shown in Figure. The stiffness of the isolator is K=KS-KN where KS is the spring stiffness and KN is the magnitude of a negative stiffness which is a function of the length of the bars and the load P. The isolator stiffness can be made to approach zero while the spring supports the weight W.

Horizontal-motion Isolation

A horizontal-motion isolator consisting of two beam-columns is illustrated in Figure. Each beam-column behaves like two fixed-free beam columns loaded axially by a weight load W. Without the weight load the beam-columns have horizontal stiffness KS With the weight load the lateral bending stiffness is reduced by the "beam-column" effect. This behavior is equivalent to a horizontal spring combined with an NSM so that the horizontal stiffness is K=KS-KN, and KN is the magnitude of the beam-column effect. Horizontal stiffness can be made to approach zero by loading the beam-columns to approach their critical buckling load.

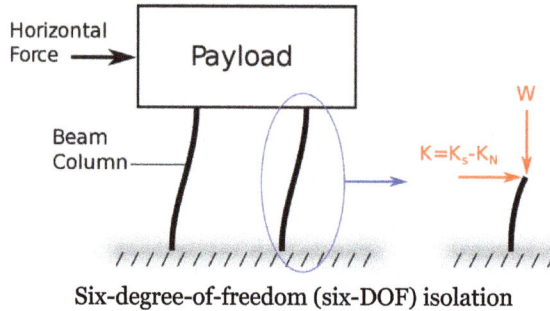

Six-degree-of-freedom (six-DOF) isolation

A six-DOF NSM isolator typically uses three isolators stacked in series: a tilt-motion isolator on top of a horizontal-motion isolator on top of a vertical-motion isolator. Figure shows a schematic of a vibration isolation system consisting of a weighted platform supported by a single six-DOF isolator incorporating the isolators of Figures. Flexures are used in place of the hinged bars shown in Figure. A tilt flexure serves as the tilt-motion isolator. A vertical-stiffness adjustment screw is used to adjust the compression force on the negative-stiffness flexures thereby changing the vertical stiffness. A vertical load adjustment screw is used to adjust for varying weight loads by raising or lowering the base of the support spring to keep the flexures in their straight, unbent operating positions.

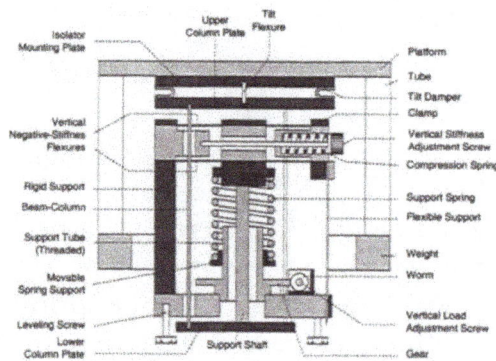

Figure 3 Schematic of Six-DOF Single-Isolator System

Vibration Isolation of Supporting Joint

The equipment or other mechanical components are necessarily linked to surrounding objects (the supporting joint - with the support; the unsupporting joint - the pipe duct or cable), thus presenting the opportunity for unwanted transmission of vibrations. Using a suitably designed vibration-isolator (absorber), vibration isolation of the supporting joint is realized. The accompanying illustration shows the attenuation of vibration levels, as measured before installation of the functioning gear on a vibration isolator as well as after installation, for a wide range of frequencies.

The Vibration Isolator

This is defined as a device that reflects and absorbs waves of oscillatory energy, extending from a piece of working machinery or electrical equipment, and with the desired effect being vibration insulation. The goal is to establish vibration isolation between a body transferring mechanical fluctuations and a supporting body (for example, between the machine and the foundation). The illustration shows a vibration isolator from the series «ВИ» (~"VI" in Roman characters), as used in shipbuilding in Russia, for example the submarine "St.Petersburg" (Lada). The depicted «ВИ» devices allow loadings ranging from 5, 40 and 300 kg. They differ in their physical sizes, but all share the same fundamental design. The structure consists of a rubber envelope that is internally reinforced by a spring. During manufacture, the rubber and the spring are intimately and permanently connected as a result of the vulcanization process that is integral to the processing of the crude rubber material. Under action of weight loading of the machine, the rubber envelope deforms, and the spring is compressed or stretched. Therefore, in the direction of the spring's cross section, twisting of the enveloping rubber occurs. The resulting elastic deformation of the rubber envelope results in very effective absorption of the vibration. This absorption is crucial to reliable vibration insulation, because it averts the potential for resonance effects. The amount of elastic deformation of the rubber largely dictates the magnitude of vibration absorption that can be attained; the entire device (including the spring itself) must be designed with this in mind. The design of the vibration isolator must also take into account potential exposure to shock loadings, in addition to the routine everyday vibrations. Lastly, the vibration isolator must also be designed for long-term durability as well as convenient integration into the environment in which it is to be used. Sleeves and flanges are typically employed in order to enable the vibration isolator to be securely fastened to the equipment and the supporting foundation.

Vibration Isolation of Unsupporting Joint

Vibration isolation of unsupporting joint is realized in the device named branch pipe a of isolating vibration.

Branch Pipe a of Isolating Vibration

Branch pipe a of isolating vibration is a part of a tube with elastic walls for reflection and absorption of waves of the oscillatory energy extending from the working pump over wall of the pipe duct. Is established between the pump and the pipe duct. On an illustration is presented the image a vibration-isolating branch pipe of a series «ВИПБ». In a structure is used the rubber envelope, which is reinforced by a spring. Properties of an envelope are similar envelope to an isolator vibration. Has the device reducing axial effort from action of internal pressure up to zero.

Subframe Isolation

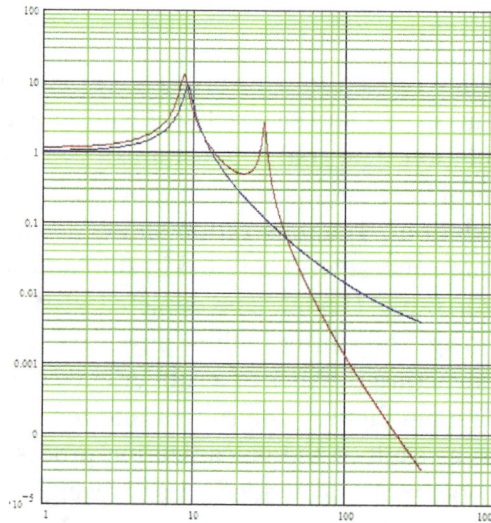

Subframe vibration isolation graph: force transmission on suspended body vs. frequency for rigidly and compliantly mounted subframes.

Another technique used to increase isolation is to use an isolated subframe. This splits the system with an additional mass/spring/damper system. This doubles the high frequency attenuation rolloff, at the cost of introducing additional low frequency modes which may cause the low frequency behaviour to deteriorate. This is commonly used in the rear suspensions of cars with Independent Rear Suspension (IRS), and in the front subframes of some cars. The graph shows the force into the body for a subframe that is rigidly bolted to the body compared with the red curve that shows a compliantly mounted subframe. Above 42 Hz the compliantly mounted subframe is superior, but below that frequency the bolted in subframe is better.

Active Isolation

Active vibration isolation systems contain, along with the spring, a feedback circuit which consists of a sensor (for example a piezoelectric accelerometer or a geophone), a controller, and an actuator. The acceleration (vibration) signal is processed by a control circuit and amplifier. Then it feeds the electromagnetic actuator, which amplifies the signal. As a result of such a feedback system, a considerably stronger suppression of vibrations is achieved compared to ordinary damping. Active isolation today is used for applications where structures smaller than a micrometer have to be produced or measured. A couple of companies produce active isolation products as OEM for research, metrology, lithography and medical systems. Another important application is the semiconductor industry. In the microchip production, the smallest structures today are below 20 nm, so the machines which produce and check them have to oscillate much less.

Seismic Retrofit

Seismic retrofitting is the modification of existing structures to make them more resistant to seismic activity, ground motion, or soil failure due to earthquakes. With better understanding of seismic demand on structures and with our recent experiences with large earthquakes near urban centers, the need of seismic retrofitting is well acknowledged. Prior to the introduction of modern seismic codes in the late 1960s for developed countries (US, Japan etc.) and late 1970s for many other parts of the world (Turkey, China etc.), many structures were designed without adequate detailing and reinforcement for seismic protection. In view of the imminent problem, various research work has been carried out. State-of-the-art technical guidelines for seismic assessment, retrofit and rehabilitation have been published around the world - such as the ASCE-SEI 41 and the New Zealand Society for Earthquake Engineering (NZSEE)'s guidelines. These codes must be regularly updated; the 1994 Northridge earthquake brought to light the brittleness of welded steel frames, for example.

Infill shear trusses — University of California dormitory, Berkeley

The retrofit techniques outlined here are also applicable for other natural hazards such as tropical cyclones, tornadoes, and severe winds from thunderstorms. Whilst current practice of seismic retrofitting is predominantly concerned with structural improvements to reduce the seismic hazard of using the structures, it is similarly essential to reduce the hazards and losses from non-structural elements. It is also important to keep in mind that there is no such thing as an earthquake-proof structure, although seismic performance can be greatly enhanced through proper initial design or subsequent modifications.

External bracing of an existing reinforced concrete parking garage (Berkeley)

Port Authority Bus Terminal

Strategies

Seismic retrofit (or rehabilitation) strategies have been developed in the past few decades following the introduction of new seismic provisions and the availability of advanced materials (e.g. fiber-reinforced polymers (FRP), fiber reinforced concrete and high strength steel). Retrofit strategies are different from retrofit techniques, where the former is the basic approach to achieve an overall retrofit performance objective, such as increasing strength, increasing deformability, reducing deformation demands while the latter is the technical methods to achieve that strategy, for example FRP jacketing.

- Increasing the global capacity (strengthening). This is typically done by the addition of cross braces or new structural walls.

- Reduction of the seismic demand by means of supplementary damping and/or use of base isolation systems.

- Increasing the local capacity of structural elements. This strategy recognises the inherent capacity within the existing structures, and therefore adopt a more cost-effective approach to selectively upgrade local capacity (deformation/ductility, strength or stiffness) of individual structural components.

- Selective weakening retrofit. This is a counter intuitive strategy to change the inelastic mechanism of the structure, while recognising the inherent capacity of the structure.

- Allowing sliding connections such as passageway bridges to accommodate additional movement between seismically independent structures.

Performance Objectives

In the past, seismic retrofit was primarily applied to achieve public safety, with engineering solutions limited by economic and political considerations. However, with the development of Performance based earthquake engineering (PBEE), several levels of performance objectives are gradually recognised:

- Public safety only. The goal is to protect human life, ensuring that the structure will not collapse upon its occupants or passersby, and that the structure can be safely exited. Under severe seismic conditions the structure may be a total economic write-off, requiring tear-down and replacement.

- Structure survivability. The goal is that the structure, while remaining safe for exit, may require extensive repair (but not replacement) before it is generally useful or considered safe for occupation. This is typically the lowest level of retrofit applied to bridges.

- Structure functionality. Primary structure undamaged and the structure is undiminished in utility for its primary application. A high level of retrofit, this ensures that any required repairs are only "cosmetic" - for example, minor cracks in plaster, drywall and stucco. This is the minimum acceptable level of retrofit for hospitals.

- Structure unaffected. This level of retrofit is preferred for historic structures of high cultural significance.

Techniques

One of many "earthquake bolts" found throughout period houses in the city of Charleston subsequent to the Charleston earthquake of 1886. They could be tightened and loosened to support the house without having to otherwise demolish the house due to instability. The bolts were directly loosely connected to the supporting frame of the house.

Common seismic retrofitting techniques fall into several categories:

External Post-tensioning

The use of external post-tensioning for new structural systems have been developed in the past decade. Under the PRESS (Precast Seismic Structural Systems), a large-scale U.S./Japan joint research program, unbonded post-tensioning high strength steel tendons have been used to achieve a moment-resisting system that has self-centering capacity. An extension of the same idea for seismic retrofitting has been experimentally tested for seismic retrofit of California bridges under a Caltrans research project and for seismic retrofit of non-ductile reinforced concrete frames. Pre-stressing can increase the capacity of structural elements such as beam, column and beam-column joints. It should be noted that external pre-stressing has been used for structural upgrade for gravity/live loading since the 1970s.

Base Isolators

Base isolation is a collection of structural elements of a building that should substantially decouple the building's structure from the shaking ground thus protecting the building's integrity and enhancing its seismic performance. This earthquake engineering technology, which is a kind of seismic vibration control, can be applied both to a newly designed building and to seismic upgrading of existing structures. Normally, excavations are made around the building and the building is separated from the foundations. Steel or reinforced concrete beams replace the connections to the foundations, while under these, the isolating pads, or base isolators, replace the material removed. While the base isolation tends to restrict transmission of the ground motion to the building, it also keeps the building positioned properly over the foundation. Careful attention to detail is required where the building interfaces with the ground, especially at entrances, stairways and ramps, to ensure sufficient relative motion of those structural elements.

Supplementary Dampers

Supplementary dampers absorb the energy of motion and convert it to heat, thus "damping" resonant effects in structures that are rigidly attached to the ground. In addition to adding energy dissipation capacity to the structure, supplementary damping can reduce the displacement and acceleration demand within the structures. In some cases, the threat of damage does not come from the initial shock itself, but rather from the periodic resonant motion of the structure that repeated ground motion induces. In the practical sense, supplementary dampers act similarly to Shock absorbers used in automotive suspensions.

Tuned Mass Dampers

Tuned mass dampers (TMD) employ movable weights on some sort of springs. These are typically employed to reduce wind sway in very tall, light buildings. Similar designs may be employed to impart earthquake resistance in eight to ten story buildings that are prone to destructive earthquake induced resonances.

Slosh Tank

A slosh tank is a large tank of fluid placed on an upper floor. During a seismic event, the fluid in this tank will slosh back and forth, but is directed by baffles - partitions that prevent the tank itself becoming resonant; through its mass the water may change or counter the resonant period of the building. Additional kinetic energy can be converted to heat by the baffles and is dissipated through the water - any temperature rise will be insignificant.

Active Control System

Very tall buildings ("skyscrapers"), when built using modern lightweight materials, might sway uncomfortably (but not dangerously) in certain wind conditions. A solution to this problem is to include at some upper story a large mass, constrained, but free to move within a limited range, and moving on some sort of bearing system such as an air cushion or hydraulic film. Hydraulic pistons, powered by electric pumps and accumulators, are actively driven to counter the wind forces and natural resonances. These may also, if properly designed, be effective in controlling excessive motion - with or without applied power - in an earthquake. In general, though, modern steel frame high rise buildings are not as subject to dangerous motion as are medium rise (eight to ten story) buildings, as the resonant period of a tall and massive building is longer than the approximately one second shocks applied by an earthquake.

Adhoc Addition of Structural Support/Reinforcement

The most common form of seismic retrofit to lower buildings is adding strength to the existing structure to resist seismic forces. The strengthening may be limited to connections between existing building elements or it may involve adding primary resisting elements such as walls or frames, particularly in the lower stories.

Connections between Buildings and their Expansion Additions

Frequently, building additions will not be strongly connected to the existing structure, but simply placed adjacent to it, with only minor continuity in flooring, siding, and roofing. As a result, the addition may have a different resonant period than the original structure, and they may easily detach from one another. The relative motion will then cause the two parts to collide, causing severe structural damage. Seismic modification will either tie the two building components rigidly together so that they behave as a single mass or it will employ dampers to expend the energy from relative motion, with appropriate allowance for this motion, such as increased spacing and sliding bridges between sections.

Exterior Reinforcement of Building

Exterior Concrete Columns

Historic buildings, made of unreinforced masonry, may have culturally important interior detailing or murals that should not be disturbed. In this case, the solution may be to add a number of steel, reinforced concrete, or poststressed concrete columns to the exterior. Careful attention must be paid to the connections with other members such as footings, top plates, and roof trusses.

Infill Shear Trusses

Shown here is an exterior shear reinforcement of a conventional reinforced concrete dormitory building. In this case, there was sufficient vertical strength in the building columns and sufficient shear strength in the lower stories that only limited shear reinforcement was required to make it earthquake resistant for this location near the Hayward fault.

Massive Exterior Structure

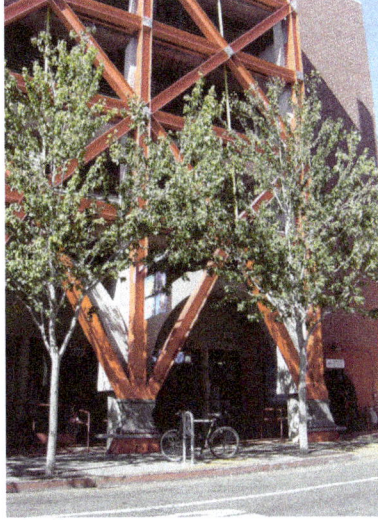

In other circumstances, far greater reinforcement is required. In the structure shown at right — a parking garage over shops — the placement, detailing, and painting of the reinforcement becomes itself an architectural embellishment.

Typical Retrofit Scenario & Solution

Soft-story Failure

Partial failure due to inadequate shear structure at garage level. Damage in San Francisco due to the Loma Prieta event.

This collapse mode is known as *soft story collapse.* In many buildings the ground level is designed for different uses than the upper levels. Low rise residential structures may be built over a parking garage which have large doors on one side. Hotels may have a tall ground floors to allow for a grand entrance or ballrooms. Office buildings may have stores in the ground floor which desire continuous windows for display.

Traditional seismic design assumes that the lower stories of a building are stronger than the upper stories and where this is not the case—if the lower story is less strong than the upper structure—the structure will not respond to earthquakes in the expected fashion. Using modern design methods, it is possible to take a weak story into account. Several failures of this type in one large apartment complex caused most of the fatalities in the 1994 Northridge earthquake.

Typically, where this type of problem is found, the weak story is reinforced to make it stronger than the floors above by adding shear walls or moment frames. Moment frames consisting of inverted U bents are useful in preserving lower story garage access, while a lower cost solution may be to use shear walls or trusses in several locations, which partially reduce the usefulness for automobile parking but still allow the space to be used for other storage.

Beam-column Joint Connections

Corner joint steel reinforcement and high tensile strength rods with grouted anti-burst jacket below

Beam-column joint connections are a common structural weakness in dealing with seismic retrofitting. Prior to the introduction of modern seismic codes in early 1970s, beam-column joints were typically non-engineered or designed. Laboratory testings have confirmed the seismic vulnerability of these poorly detailed and under-designed connections. Failure of beam-column joint connections can typically lead to catastrophic collapse of a frame-building, as often observed in recent earthquakes

For reinforced concrete beam-column joints - various retrofit solutions have been proposed and tested in the past 20 years. Philosophically, the various seismic retrofit strategies discussed above can be implemented for reinforced concrete joints. Concrete or steel jacketing have been a popular retrofit technique until the advent of composite materials such as Carbon fiber-reinforced polymer (FRP). Composite materials such as carbon FRP and aramic FRP have been extensively tested for use in seismic retrofit with some success. One novel technique includes the use of selective weakening of the beam and added external post-tensioning to the joint in order to achieve flexural hinging in the beam, which is more desirable in terms of seismic design.

Widespread weld failures at beam-column joints of low-to-medium rise steel buildings during the Northridge 1994 earthquake for example, have shown the structural defiencies of these 'modern-designed' post-1970s welded moment-resisting connections. A subsequent SAC research project has documented, tested and proposed several retrofit solutions for these welded steel moment-resisting connections. Various retrofit solutions have been developed for these welded joints - such as a) weld strengthening and b) addition of steel haunch or 'dog-bone' shape flange.

Following the Northridge earthquake, a number of steel moment -frame buildings were found to have experienced brittle fractures of beam to column connections. Discovery of these unanticipated brittle fractures of framing connections was alarming to engineers and the building industry. Starting in the 1960s, engineers began to regard welded steel moment-frame buildings as being among the most ductile systems contained in the building code. Many engineers believed that steel moment-frame buildings were essentially invulnerable to earthquake induced damage and thought that should damage occur, it would be limited to ductile yielding of members and connections. Observation of damage sustained by buildings in the 1994 Northridge earthquake indicated that contrary to the intended behavior, in many cases, brittle fractures initiated within the connections at very low levels of plastic demand. In September, 1994, The SAC joint Venture, AISC, AISI, and NIST jointly convened an international workshop in Los Angeles to coordinate the efforts of various participants and to lay the foundation for systematic investigation and resolution of the problem. In September 1995 the SAC Joint Venture entered into a contractual agreement with FEMA to conduct Phase II of the SAC Steel project. Under Phase II, SAC continued its extensive problem-focused study of the performance of moment resisting steel frames and connections of various configurations, with the ultimate goal of developing seismic design criteria for steel construction. As a result of these studies it is now known that the typical moment-resisting connection detail employed in steel moment frame construction prior to the 1994 Northridge earthquake had a number of features that rendered it inherently susceptible to brittle fracture.

Shear Failure within Floor Diaphragm

Floors in wooden buildings are usually constructed upon relatively deep spans of wood, called joists, covered with a diagonal wood planking or plywood to form a subfloor upon which the finish floor surface is laid. In many structures these are all aligned in the same direction. To prevent the beams from tipping over onto their side, blocking is used at each end, and for additional stiffness, blocking or diagonal wood or metal bracing may be placed between beams at one or more points in their spans. At the outer edge it is typical to use a single depth of blocking and a perimeter beam overall.

If the blocking or nailing is inadequate, each beam can be laid flat by the shear forces applied to the building. In this position they lack most of their original strength and the structure may further collapse. As part of a retrofit the blocking may be doubled, especially at the outer edges of the building. It may be appropriate to add additional nails between the sill plate of the perimeter wall erected upon the floor diaphragm, although this will require exposing the sill plate by removing interior plaster or exterior siding. As the sill plate may be quite old and dry and substantial nails must be used, it may be necessary to pre-drill a hole for the nail in the old wood to avoid splitting. When the wall is opened for this purpose it may also be appropriate to tie vertical wall elements into the foundation using specialty connectors and bolts glued with epoxy cement into holes drilled in the foundation.

Sliding off Foundation and "Cripple Wall" Failure

House slid off of foundation

Low cripple wall collapse and detachment of structure from concrete stairway

Single or two story wood-frame domestic structures built on a perimeter or slab foundation are relatively safe in an earthquake, but in many structures built before 1950 the sill plate that sits between the concrete foundation and the floor diaphragm (perimeter foundation) or studwall (slab foundation) may not be sufficiently bolted in. Additionally, older attachments (without substantial corrosion-proofing) may have corroded to a point of weakness. A sideways shock can slide the building entirely off of the foundations or slab.

Often such buildings, especially if constructed on a moderate slope, are erected on a platform connected to a perimeter foundation through low stud-walls called "cripple wall" or *pin-up*. This low wall structure itself may fail in shear or in its connections to itself at the corners, leading to the building moving diagonally and collapsing the low walls. The likelihood of failure of the pin-up can be reduced by ensuring that the corners are well reinforced in shear and that the shear panels are well connected to each other through the corner posts. This requires structural grade sheet plywood, often treated for rot resistance. This grade of plywood is made without interior unfilled knots and with more, thinner layers than common plywood. New buildings designed to resist earthquakes will typically use OSB (oriented strand board), sometimes with metal joins between panels, and with well attached stucco covering to enhance its performance. In many modern tract homes, especially those built upon expansive (clay) soil the building is constructed upon a single and relatively thick monolithic slab, kept in one piece by high tensile rods that are stressed after the slab has set. This poststressing places the concrete under compression - a condition under which it is extremely strong in bending and so will not crack under adverse soil conditions.

Multiple Piers in Shallow Pits

Some older low-cost structures are elevated on tapered concrete pylons set into shallow pits, a method frequently used to attach outdoor decks to existing buildings. This is seen in conditions of damp soil, especially in tropical conditions, as it leaves a dry ventilated space under the house, and in far northern conditions of permafrost (frozen mud) as it keeps the building's warmth from destabilizing the ground beneath. During an earthquake, the pylons may tip, spilling the building to the ground. This can be overcome by using deep-bored holes to contain cast-in-place reinforced pylons, which are then secured to the floor panel at the corners of the building. Another technique is to add sufficient diagonal bracing or sections of concrete shear wall between pylons.

Reinforced Concrete Column Burst

Jacketed and grouted column on left, unmodified on right

Reinforced concrete columns typically contain large diameter vertical rebar (reinforcing bars) arranged in a ring, surrounded by lighter-gauge hoops of rebar. Upon analysis of failures due to earthquakes, it has been realized that the weakness was not in the vertical bars, but rather in inadequate strength and quantity of hoops. Once the integrity of the hoops is breached, the vertical rebar can flex outward, stressing the central column of concrete. The concrete then simply crumbles into small pieces, now unconstrained by the surrounding rebar. In new construction a greater amount of hoop-like structures are used.

One simple retrofit is to surround the column with a jacket of steel plates formed and welded into a single cylinder. The space between the jacket and the column is then filled with concrete, a process called grouting. Where soil or structure conditions require such additional modification, additional pilings may be driven near the column base and concrete pads linking the pilings to the pylon are fabricated at or below ground level. In the example shown not all columns needed to be modified to gain sufficient seismic resistance for the conditions expected. (This location is about a mile from the Hayward Fault Zone.)

Reinforced Concrete Wall Burst

Concrete walls are often used at the transition between elevated road fill and overpass structures. The wall is used both to retain the soil and so enable the use of a shorter span and also to transfer the weight of the span directly downward to footings in undisturbed soil. If these walls are inadequate they may crumble under the stress of an earthquake's induced ground motion.

One form of retrofit is to drill numerous holes into the surface of the wall, and secure short L-shaped sections of rebar to the surface of each hole with epoxy adhesive. Additional vertical and horizontal rebar is then secured to the new elements, a form is erected, and an additional layer of concrete is poured. This modification may be combined with additional footings in excavated trenches and additional support ledgers and tiebacks to retain the span on the bounding walls.

Brick Wall Resin and Glass Fiber Reinforcement

Brick building structures have been reinforced with coatings of glass fiber and appropriate resin (epoxy or polyester). In lower floors these may be applied over entire exposed surfaces, while in upper floors this may be confined to narrow areas around window and door openings. This application provides tensile strength that stiffens the wall against bending away from the side with the application. The efficient protection of an entire building requires extensive analysis and engineering to determine the appropriate locations to be treated.

Lift

Where moist or poorly consolidated alluvial soil interfaces in a "beach like" structure against underlying firm material, seismic waves traveling through the alluvium can be amplified, just as are water waves against a sloping beach. In these special conditions, vertical accelerations up to twice the force of gravity have been measured. If a building is not secured to a well-embedded foundation it is possible for the building to be thrust from (or with) its foundations into the air, usually with severe damage upon landing. Even if it is well-founded, higher portions such as upper stories or roof structures or attached structures such as canopies and porches may become detached from the primary structure.

Good practices in modern, earthquake-resistant structures dictate that there be good vertical connections throughout every component of the building, from undisturbed or engineered earth to foundation to sill plate to vertical studs to plate cap through each floor and continuing to the roof structure. Above the foundation and sill plate the connections are typically made using steel strap or sheet stampings, nailed to wood members using special hardened high-shear strength nails, and heavy angle stampings secured with through bolts, using large washers to prevent pull-through. Where inadequate bolts are provided between the sill plates and a foundation in existing construction (or are not trusted due to possible corrosion), special clamp plates may be added, each of which is secured to the foundation using expansion bolts inserted into holes drilled in an exposed face of concrete. Other members must then be secured to the sill plates with additional fittings.

Soil

One of the most difficult retrofits is that required to prevent damage due to soil failure. Soil failure can occur on a slope, a slope failure or landslide, or in a flat area due to liquefaction of water-saturated sand and/or mud. Generally, deep pilings must be driven into stable soil (typically hard mud or sand) or to underlying bedrock or the slope must be stabilized. For buildings built atop previous landslides the practicality of retrofit may be limited by economic factors, as it is not practical to stabilize a large, deep landslide. The likelihood of landslide or soil failure may also depend upon seasonal factors, as the soil may be more stable at the beginning of a wet season than at the beginning of the dry season. Such a "two season" *Mediterranean climate* is seen throughout California.

In some cases, the best that can be done is to reduce the entrance of water runoff from higher, stable elevations by capturing and bypassing through channels or pipes, and to drain water infiltrated directly and from subsurface springs by inserting horizontal perforated tubes. There are numerous locations in California where extensive developments have been built atop archaic landslides, which have not moved in historic times but which (if both water-saturated and shaken by an earthquake) have a high probability of moving *en masse*, carrying entire sections of suburban development to new locations. While the most modern of house structures (well tied to monolithic concrete foundation slabs reinforced with post tensioning cables) may survive such movement largely intact, the building will no longer be in its proper location.

Utility Pipes and Cables: Risks

Natural gas and propane supply pipes to structures often prove especially dangerous during and after earthquakes. Should a building move from its foundation or fall due to cripple wall collapse, the ductile iron pipes transporting the gas within the structure may be broken, typically at the location of threaded joints. The gas may then still be provided to the pressure regulator from higher pressure lines and so continue to flow in substantial quantities; it may then be ignited by a nearby source such as a lit pilot light or arcing electrical connection.

There are two primary methods of automatically restraining the flow of gas after an earthquake, installed on the low pressure side of the regulator, and usually downstream of the gas meter.

- A caged metal ball may be arranged at the edge of an orifice. Upon seismic shock, the ball will roll into the orifice, sealing it to prevent gas flow. The ball may later be reset by the use of an external magnet. This device will respond only to ground motion.

- A flow-sensitive device may be used to close a valve if the flow of gas exceeds a set threshold (very much like an electrical circuit breaker). This device will operate independently of seismic motion, but will not respond to minor leaks which may be caused by an earthquake.

It appears that the most secure configuration would be to use one of each of these devices in series.

Tunnels

Unless the tunnel penetrates a fault likely to slip, the greatest danger to tunnels is a landslide blocking an entrance. Additional protection around the entrance may be applied to divert any falling material (similar as is done to divert snow avalanches) or the slope above the tunnel may be stabilized in some way. Where only small- to medium-sized rocks and boulders are expected to fall, the entire slope may be covered with wire mesh, pinned down to the slope with metal rods. This is also a common modification to highway cuts where appropriate conditions exist.

Underwater Tubes

The safety of underwater tubes is highly dependent upon the soil conditions through which the tunnel was constructed, the materials and reinforcements used, and the maximum predicted earthquake expected, and other factors, some of which may remain unknown under current knowledge.

BART Tube

A tube of particular structural, seismic, economic, and political interest is the BART (Bay Area Rapid Transit) transbay tube. This tube was constructed at the bottom of San

Francisco Bay through an innovative process. Rather than pushing a shield through the soft bay mud, the tube was constructed on land in sections. Each section consisted of two inner train tunnels of circular cross section, a central access tunnel of rectangular cross section, and an outer oval shell encompassing the three inner tubes. The intervening space was filled with concrete. At the bottom of the bay a trench was excavated and a flat bed of crushed stone prepared to receive the tube sections. The sections were then floated into place and sunk, then joined with bolted connections to previously-placed sections. An overfill was then placed atop the tube to hold it down. Once completed from San Francisco to Oakland, the tracks and electrical components were installed. The predicted response of the tube during a major earthquake was likened to be as that of a string of (cooked) spaghetti in a bowl of gelatin dessert. To avoid overstressing the tube due to differential movements at each end, a sliding slip joint was included at the San Francisco terminus under the landmark Ferry Building.

The engineers of the construction consortium PBTB (Parsons Brinckerhoff-Tudor-Bechtel) used the best estimates of ground motion available at the time, now known to be insufficient given modern computational analysis methods and geotechnical knowledge. Unexpected settlement of the tube has reduced the amount of slip that can be accommodated without failure. These factors have resulted in the slip joint being designed too short to ensure survival of the tube under possible (perhaps even likely) large earthquakes in the region. To correct this deficiency the slip joint must be extended to allow for additional movement, a modification expected to be both expensive and technically and logistically difficult. Other retrofits to the BART tube include vibratory consolidation of the tube's overfill to avoid potential liquefying of the overfill, which has now been completed. (Should the overfill fail there is a danger of portions of the tube rising from the bottom, an event which could potentially cause failure of the section connections.)

Bridge Retrofit

Bridges have Several Failure Modes.

Expansion Rockers

Many short bridge spans are statically anchored at one end and attached to rockers at the other. This rocker gives vertical and transverse support while allowing the bridge span to expand and contract with temperature changes. The change in the length of the span is accommodated over a gap in the roadway by comb-like expansion joints. During severe ground motion, the rockers may jump from their tracks or be moved beyond their design limits, causing the bridge to unship from its resting point and then either become misaligned or fail completely. Motion can be constrained by adding ductile or high-strength steel restraints that are friction-clamped to beams and designed to slide under extreme stress while still limiting the motion relative to the anchorage.

Deck Rigidity

Additional diagonals were inserted under both decks of this bridge

Suspension bridges may respond to earthquakes with a side-to-side motion exceeding that which was designed for wind gust response. Such motion can cause fragmentation of the road surface, damage to bearings, and plastic deformation or breakage of components. Devices such as hydraulic dampers or clamped sliding connections and additional diagonal reenforcement may be added.

Lattice Girders, Beams, and Ties

Obsolete riveted lattice members

Lattice girders consist of two "I"-beams connected with a criss-cross lattice of flat strap or angle stock. These can be greatly strengthened by replacing the open lattice with plate members. This is usually done in concert with the replacement of hot rivets with bolts.

Bolted plate lattice replacement, forming box members

Hot Rivets

Many older structures were fabricated by inserting red-hot rivets into pre-drilled holes; the soft rivets are then peened using an air hammer on one side and a bucking bar on the head end. As these cool slowly, they are left in an annealed (soft) condition, while the plate, having been hot rolled and quenched during manufacture, remains relatively hard. Under extreme stress the hard plates can shear the soft rivets, resulting in failure of the joint.

The solution is to burn out each rivet with an oxygen torch. The hole is then prepared to a precise diameter with a reamer. A special *locator bolt*, consisting of a head, a shaft matching the reamed hole, and a threaded end is inserted and retained with a nut, then tightened with a wrench. As the bolt has been formed from an appropriate high-strength alloy and has also been heat-treated, it is not subject to either the plastic shear failure typical of hot rivets nor the brittle fracture of ordinary bolts. Any partial failure will be in the plastic flow of the metal secured by the bolt; with proper engineering any such failure should be non-catastrophic.

Fill and Overpass

Elevated roadways are typically built on sections of elevated earth fill connected with bridge-like segments, often supported with vertical columns. If the soil fails where a bridge terminates, the bridge may become disconnected from the rest of the roadway and break away. The retrofit for this is to add additional reinforcement to any support-ing wall, or to add deep caissons adjacent to the edge at each end and connect them with a supporting beam under the bridge.

Another failure occurs when the fill at each end moves (through resonant effects) in bulk, in opposite directions. If there is an insufficient founding shelf for the overpass, then it may fall. Additional shelf and ductile stays may be added to attach the overpass to the footings at one or both ends. The stays, rather than being fixed to the beams, may instead be clamped to them. Under moderate loading, these keep the overpass centered in the gap so that it is less likely to slide off its founding shelf at one end. The ability for the fixed ends to slide, rather than break, will prevent the complete drop of the struc-ture if it should fail to remain on the footings.

Viaducts

Large sections of roadway may consist entirely of viaduct, sections with no con-nection to the earth other than through vertical columns. When concrete columns are used, the detailing is critical. Typical failure may be in the toppling of a row of columns due either to soil connection failure or to insufficient cylindrical wrap-ping with rebar. Both failures were seen in the 1995 Great Hanshin earthquake in Kobe, Japan, where an entire viaduct, centrally supported by a single row of large columns, was laid down to one side. Such columns are reinforced by excavating to the foundation pad, driving additional pilings, and adding a new, larger pad, well connected with rebar alongside or into the column. A column with insufficient wrapping bar, which is prone to burst and then hinge at the bursting point, may be completely encased in a circular or elliptical jacket of welded steel sheet and grouted as described above.

Cypress Freeway viaduct collapse. Note failure of inadequate anti-burst wrapping and lack of connection between upper and lower vertical elements.

Sometimes viaducts may fail in the connections between components. This was seen in the failure of the Cypress Freeway in Oakland, California, during the Loma Prieta earthquake. This viaduct was a two-level structure, and the upper portions of the columns were not well connected to the lower portions that supported the lower level; this caused the upper deck to collapse upon the lower deck. Weak connections such as these require additional external jacketing - either through external steel components or by a complete jacket of reinforced concrete, often using stub connections that are glued (using epoxy adhesive) into numerous drilled holes. These stubs are then connected to additional wrappings, external forms (which may be temporary or permanent) are erected, and additional concrete is poured into the space. Large connected structures similar to the Cypress Viaduct must also be properly analyzed in their entirety using dynamic computer simulations.

Residential Retrofit

Side-to-side forces cause most earthquake damage. Bolting of the mudsill to the foundation and application of plywood to cripple walls are a few basic retrofit techniques which homeowners may apply to wood-framed residential structures to mitigate the effects of seismic activity. The City of San Leandro created guidelines for these procedures, as outlined in the following pamphlet. Public awareness and initiative are critical to the retrofit and preservation of existing building stock, and such efforts as those of the Association of Bay Area Governments are instrumental in providing informational resources to seismically active communities.

Wood Frame Structure

Most houses in North America are wood-framed structures. Wood is one of the best materials for earthquake-resistant construction since it is lightweight and more flexible than masonry. It is easy to work with and less expensive than steel, masonry, or concrete. In older homes the most significant weaknesses are the connection from the wood-framed walls to the foundation and the relatively weak "cripple-walls." (Cripple walls are the short wood walls that extend from the top of the foundation to the lowest floor level in houses that have raised floors.) Adding connections from the base of the wood-framed structure to the foundation is almost always an important part of a seismic retrofit. Bracing the cripple-walls to resist side-to-side forces is essential in houses with cripple walls; bracing is usually done with plywood. Oriented strand board (OSB) does not perform as consistently as plywood, and is not the favored choice of retrofit designers or installers.

Retrofit methods in older woodframe structures may consist of the following, and other methods not described here.

- The lowest plate rails of walls (usually called "mudsills" or "foundation sills" in North America) are bolted to a continuous foundation, or secured with rigid metal connectors bolted to the foundation so as to resist side-to-side forces.

- *Cripple walls* are braced with plywood.

- Selected vertical elements (typically the posts at the ends of plywood wall bracing panels) are connected to the foundation. These connections are intended to prevent the braced walls from rocking up and down when subjected to back-and-forth forces at the top of the braced walls, not to resist the wall or house "jumping" off the foundation (which almost never occurs).

- In two story buildings using "platform framing" (sometimes called "western" style construction, where walls are progressively erected upon the lower story's upper diaphragm, unlike "eastern" or *balloon framing*), the upper walls are connected to the lower walls with tension elements. In some cases, connections may be extended vertically to include retention of certain roof elements. This sort of strengthening is usually very costly with respect to the strength gained.

- Vertical posts are secured to the beams or other members they support. This is particularly important where loss of support would lead to collapse of a segment of a building. Connections from posts to beams cannot resist appreciable side-to-side forces; it is much more important to strengthen around the perimeter of a building (bracing the cripple-walls and supplementing foundation-to-wood-framing connections) than it is to reinforce post-to-beam connections.

Wooden framing is efficient when combined with masonry, if the structure is properly designed. In Turkey, the traditional houses (bagdadi) are made with this technology. In El Salvador, wood and bamboo are used for residential construction.

Reinforced and Unreinforced Masonry

In many parts of developing countries such as Pakistan, Iran and China, unreinforced or in some cases reinforced masonry is the predominantly form of structures for rural residential and dwelling. Masonry was also a common construction form in the early part of the 20th century, which implies that a substantial number of these at-risk masonry structures would have significant heritage value. Masonry walls that are not reinforced are especially hazardous. Such structures may be more appropriate for replacement than retrofit, but if the walls are the principal load bearing elements in structures of modest size they may be appropriately reinforced. It is especially important that floor and ceiling beams be securely attached to the walls. Additional vertical supports in the form of steel or reinforced concrete may be added.

In the western United States, much of what is seen as masonry is actually brick or stone veneer. Current construction rules dictate the amount of *tie–back* required, which consist of metal straps secured to vertical structural elements. These straps extend into mortar courses, securing the veneer to the primary structure. Older structures may not secure this sufficiently for seismic safety. A weakly secured veneer in a house interior (sometimes used to face a fireplace from floor to ceiling) can be especially dangerous to occupants. Older masonry chimneys are also dangerous if they have substantial vertical extension above the roof. These are prone to breakage at the roofline and may fall into the house in a single large piece. For retrofit, additional supports may be added; however, it is extremely expensive to strengthen an existing masonry chimney to conform with contemporary design standards. It is best to simply remove the extension and replace it with lighter materials, with special metal flue replacing the flue tile and a wood structure replacing the masonry. This may be matched against existing brickwork by using very thin veneer (similar to a tile, but with the appearance of a brick).

References

- Chu, S.Y.; Soong, T.T.; Reinhorn, A.M. (2005). Active, Hybrid and Semi-Active Structural Control. John Wiley & Sons. ISBN 0-470-01352-4.

- Reitherman, Robert (2012). Earthquakes and Engineers: An International History. Reston, VA: ASCE Press. ISBN 9780784410714.

- Harris, C., Piersol, A., Harris Shock and Vibration Handbook, Fifth Edition, McGraw-Hill, (2002), ISBN 0-07-137081-1.

- Reitherman, Robert (2012). Earthquakes and Engineers: An International History. Reston, VA: ASCE Press. pp. 486–487. ISBN 9780784410714.

6

Understanding Earthquake Resistant Structures

Earthquake resistant structures are built to withstand earthquakes and to minimize damage to life and property. Ancient architecture worked around this problem by constructing stiff and strong structures but with the modern advancements in architectural science, the focus has shifted to keeping functionality intact while also incorporating elements that will help structures suffer less damage. To this end the chapter explores the inclusion of steel plate shear walls, its advantages and analytical models.

Earthquake Resistant Structures

Earthquake-resistant structures are structures designed to withstand earthquakes. While no structure can be entirely immune to damage from earthquakes, the goal of earthquake-resistant construction is to erect structures that fare better during seismic activity than their conventional counterparts.

According to building codes, earthquake-resistant structures are intended to withstand the largest earthquake of a certain probability that is likely to occur at their location. This means the loss of life should be minimized by preventing collapse of the buildings for rare earthquakes while the loss of functionality should be limited for more frequent ones.

To combat earthquake destruction, the only method available to ancient architects was to build their landmark structures to last, often by making them excessively stiff and strong, like the El Castillo pyramid at Chichen Itza.

Currently, there are several design philosophies in earthquake engineering, making use of experimental results, computer simulations and observations from past earthquakes to offer the required performance for the seismic threat at the site of interest. These range from appropriately sizing the structure to be strong and ductile enough to survive the shaking with an acceptable damage, to equipping it with base isolation or using structural vibration control technologies to minimize any forces and deformations. While the former is the method typically applied in most earthquake-resistant structures, important facilities, landmarks and cultural heritage buildings use the more advanced (and expensive) techniques of isolation or control to survive strong shaking with minimal damage. Examples of such applications are the Cathedral of Our Lady of the Angels and the Acropolis Museum.

Trends and Projects

Some of the new trends and/or projects in the field of earthquake engineering structures are presented.

Building Materials

Based on experience in earthquakes in Eastern European and in Central Asian countries where precast concrete has been widely used as construction material, it can be concluded that their seismic performance has been fairly satisfactory. Based on studies in New Zealand, relating to Christchurch earthquakes, precast concrete designed and installed in accordance with modern codes performed well. According to the Earthquake Engineering Research Institute, precast panel buildings had good durability during the earthquake in Armenia, compared to precast frame-panels.

Earthquake Shelter

One Japanese construction company has developed a six-foot cubical shelter, presented as an alternative to earthquake-proofing an entire building.

Concurrent Shake-table Testing

Concurrent shake-table testing of two or more building models is a vivid, persuasive and effective way to validate earthquake engineering solutions experimentally.

Thus, two wooden houses built before adoption of the 1981 Japanese Building Code were moved to E-Defense for testing. The left house was rein-forced to enhance its seismic resistance, while the other one was not. These two models were set on E-Defense platform and tested simultaneously .

Combined Vibration Control Solution

Seismically retrofitted Municipal Services Building in Glendale, CA

Designed by architect Merrill W. Baird of Glendale, working in collaboration with A. C. Martin Architects of Los Angeles, the Municipal Services Building at 633 East Broadway, Glendale was completed in 1966 . Prominently sited at the corner of East Broadway and Glendale Avenue, this civic building serves as a heraldic element of Glendale's civic center.

In October 2004 Architectural Resources Group (ARG) was contracted by Nabih Youssef & Associates, Structural Engineers, to provide services regarding a historic resource assessment of the building due to a proposed seismic retrofit.

In 2008, the Municipal Services Building of the City of Glendale, California was seismically retrofitted using an innovative combined vibration control solution: the existing elevated building foundation of the building was put on high damping rubber bearings.

Steel Plate Walls System

Coupled steel plate shear walls, Seattle

The Ritz-Carlton/JW Marriott hotel building engaging the advanced
steel plate shear walls system, LA

A steel plate shear wall (SPSW) consists of steel infill plates bounded by a column-beam system. When such infill plates occupy each level within a framed bay of a structure, they constitute a SPSW system. Whereas most earthquake resistant construction methods are adapted from older systems, SPSW was invented entirely to withstand seismic activity.

SPSW behavior is analogous to a vertical plate girder cantilevered from its base. Similar to plate girders, the SPSW system optimizes component performance by taking advantage of the post-buckling behavior of the steel infill panels.

The Ritz-Carlton/JW Marriott hotel building, a part of the LA Live development in Los Angeles, California, is the first building in Los Angeles that uses an advanced steel plate shear wall system to resist the lateral loads of strong earthquakes and winds.

Kashiwazaki-Kariwa Nuclear Power Plant is Partially Upgraded

The Kashiwazaki-Kariwa Nuclear Power Plant, the largest nuclear generating station in the world by net electrical power rating, happened to be near the epicenter of the strongest M_w 6.6 July 2007 Chūetsu offshore earthquake. This initiated an extended shutdown for structural inspection which indicated that a greater earthquake-proofing was needed before operation could be resumed.

On May 9, 2009, one unit (Unit 7) was restarted, after the seismic upgrades. The test run had to continue for 50 days. The plant had been completely shut down for almost 22 months following the earthquake.

Seismic Test of Seven-story Building

A destructive earthquake struck a lone, wooden condominium in Japan . The experiment was webcast live on July 14, 2009 to yield insight on how to make wooden structures stronger and better able to withstand major earthquakes .

The Miki shake at the Hyogo Earthquake Engineering Research Center is the capstone experiment of the four-year NEESWood project, which receives its primary support from the U.S. National Science Foundation Network for Earthquake Engineering Simulation (NEES) Program.

"NEESWood aims to develop a new seismic design philosophy that will provide the necessary mechanisms to safely increase the height of wood-frame structures in active seismic zones of the United States, as well as mitigate earthquake damage to low-rise wood-frame structures," said Rosowsky, Department of Civil Engineering at Texas A&M University. This philosophy is based on the application of seismic damping systems for wooden buildings. The systems, which can be installed inside the walls of most wooden buildings, include strong metal frame, bracing and dampers filled with viscous fluid.

Superframe RC Earthquake Proof Structure

The proposed system is composed of core walls, hat beams incorporated into the top level, outer columns and viscous dampers vertically installed between the tips of the hat beams and the outer columns. During an earthquake,the hat beams and outer columns act as outriggers and reduce the overturning moment in the core, and the installed dampers also reduce the moment and the lateral deflection of the structure. This innovative system can eliminate inner beams and inner columns on each floor, and thereby provide buildings with column-free floor space even in highly seismic regions.

Steel Plate Shear Wall

A steel plate shear wall *(SPSW)* consists of steel infill plates bounded by boundary elements.

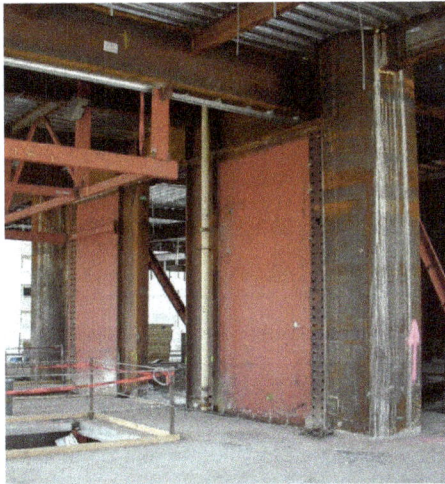

Coupled steel plate shear wall - Seattle - WA

Overview

They constitute an SPSW. Its behavior is analogous to a vertical plate girder cantilevered from its base. Similar to plate girders, the SPW system optimizes component performance by taking advantage of the post-buckling behavior of the steel infill panels. An SPW frame can be idealized as a vertical cantilever plate girder, in which the steel plates act as the web, the columns act as the flanges and the cross beams represent the transverse stiffeners. The theory that governs plate design should not be used in design of SPW structures since the relatively high bending strength and stiffness of the beams and columns have a significant effect in the post-buckling behavior.

Capacity design of structures is: to control failure in a building by pre-selecting localized ductile fuses (or weak links) to act as the primary location for energy dissipation when a building is subjected to extreme loading. The structure is designed such that all inelastic action (or damage) occurs at these critical locations (the fuses), which are designed to behave in a ductile and stable manner. Conversely, all other structural elements are protected against failure or collapse by limiting the load transfer to these elements to the yield capacity of the fuses. In SPSWs, the infill plates are meant to serve as the fuse elements. When damaged during an extreme loading event, they can be replaced at a reasonable cost and restore full integrity of the building. In general, SPWs are categorized based on their performance, selection of structural and load-bearing systems, and the presence of perforations or stiffeners (Table 1).

A significant amount of valuable research has been performed on the static and dynamic behavior of SPSWs. Much research has been conducted to not only help determine the behavior, response and performance of SPWs under cyclic and dynamic loading, but also as a means to help advance analysis and design methodologies for the engineering community.

The pioneering work of Kulak and co-investigators at the University of Alberta in Canada led to a simplified method for analyzing a thin unstiffened SPSW - the strip model. This model is incorporated in Chapter 20 of the most recent Canadian Steel Design Standard (CAN/CSA S16-01) and the National Earthquake Hazard Reduction Program (NEHRP) provisions in the US.

Table 1. Categorization of steel plate walls based on performance characteristics and expectations

Performance Characteristic	Performance Expectations or SPW Characteristics
Type of Loading carried by SPW	Lateral Load Only / Lateral Load + Wall's Dead Load (or so called 50% Gravity Load)/ Gravity + Lateral Loads
Structural System	Single wall with and without infill Columns / Coupled wall with and without infill Columns
Stiffener Spacing and Size	Post-Buckling effect can be seen in the sub panels / Panel buckles with the stiffeners globally / Stiffeners produces sub-panels which can be categorized as thick panel
Web Plate Behavior	Web plate yields before critical elastic buckling occurs (thick plate) / Web plate buckles elastically, develops post-buckling tension field, then yields (thin plate)
Web Plate Perforations	With perforations / Without perforations

History

In the past two decades the steel plate shear wall (SPSW), also known as the steel plate wall (SPW), has been used in a number of buildings in Japan and North America as part of the lateral force resisting system. In earlier days, SPSWs were treated like vertically oriented plate girders and design procedures tended to be very conservative. Web buckling was prevented through extensive stiffening or by selecting an appropriately thick web plate, until more information became available on the post-buckling characteristics of web plates. Although the plate girder theory seems appropriate for the design of an SPW structure, a very important difference is the relatively high bending strength and stiffness of the beams and columns that form the boundary elements of the wall. These members are expected to have a significant effect on the overall behaviour of a building incorporating this type of system and several researchers have focused on this aspect of SPWs. The energy dissipating qualities of the web plate under extreme cyclic loading has raised the prospect of using SPSWs as a promising alternative to conventional systems in high-risk seismic regions. A further benefit is that the diagonal tension field of the web plate acts like a diagonal brace in a braced frame and thus completes the truss action, which is known to be an efficient means to control wind drift.

Advantages

From a designer's point of view, steel plate walls have become a very attractive alternative to other steel systems, or to replace reinforced concrete elevator cores and shear walls. In comparative studies it has been shown that the overall costs of a building can be reduced significantly when considering the following advantages:

- An SPW system, when designed and detailed properly, has relatively large energy dissipation capability with stable hysteretic behaviour, thus being very attractive for high risk earthquake zones.

- Because the web tension field acts much like a diagonal brace, an SPW system has relatively high initial stiffness, and is thus very effective in limiting wind drift.

- Compared to reinforced concrete shear walls, SPWs are much lighter, which ultimately reduces the demand on columns and foundations, and reduces the seismic load, which is proportional to the mass of the structure.

- Compared to reinforced concrete construction, the erection process of an all-steel building is significantly faster, thus reducing the construction duration, which is an important factor affecting the overall cost of a project.

- By using shop-welded, field-bolted SPWs, field inspection is improved and a high level of quality control can be achieved.

- For architects, the increased versatility and space savings because of the smaller cross-section of SPWs, compared to reinforced concrete shear walls, is a distinct benefit, especially in high-rise buildings, where reinforced concrete shear walls in lower floors become very thick and occupy a large proportion of the floor plan.

- All-steel construction with SPWs is a practical and efficient solution for cold regions where concrete construction may not be feasible, as very low temperatures complicate construction and freeze-thaw cycles can result in durability problems.

- In seismic retrofit applications, SPWs are typically much easier and faster to install than reinforced concrete shear walls, which is a critical issue when building occupancy needs to be maintained throughout the construction time.

- In the event of inelastic response, steel panels are more readily replaced, and repairs are otherwise more simple than for equivalent reinforced-concrete systems.

In comparison with conventional bracing systems, steel panels have the advantage of being a redundant, continuous system exhibiting relatively stable and ductile behaviour under severe cyclic loading (Tromposch and Kulak, 1987). This benefit along with the high stiffness of the plates acting like tension braces to maintain stability, strongly qualifies the SPW as an ideal energy dissipation system in high risk seismic regions, while providing an efficient system to reduce lateral drift. Thus, some of the advantages of using SPWs compared with conventional bracing systems are as follows:

- Reduces seismic force demand due to higher SPW ductility characteristics and inherent redundancy and continuity

- Accelerates structural steel erection by using shop-welded and field-bolted steel panels, and thus, less inspection and reduced quality control costs

- Permits efficient design of lateral-resisting systems by distributing large forces evenly.

A steel plate shear element consists of steel infill plates bounded by a column-beam system. When these infill plates occupy each level within a framed bay of a structure, they constitute an SPW. Its behaviour is analogous to a vertical plate girder cantilevered from its base. Similar to plate girders, the SPW system optimizes component performance by taking advantage of the post-buckling behaviour of the steel infill panels. An SPW frame can be idealized as a vertical cantilever plate girder, in which the steel plates act as the web, the columns act as the flanges and the cross beams1 represent the transverse stiffeners. The theory that governs the design of plate girders for buildings proposed by Basler in 1960, should not be used in design of SPW structures since the relatively high bending strength and stiffness of the beams and columns is expected to have a significant effect in the post-buckling behaviour. However, Basler's theory could be used as a basis to derive an analytical model for SPW systems.

Designers pioneering the use of SPWs did not have much experience nor existing data to rely upon. Typically, web plate design failed to consider post-buckling behaviour under shear, thus ignoring the advantage of the tension field and its added benefits for drift control and shear resistance. Furthermore, the inelastic deformation capacity of this highly redundant system had not been utilized, also ignoring the significant energy dissipation capability that is of great importance for buildings in high-risk seismic zones. One of the first researchers to investigate the behaviour of SPWs more closely was Kulak at the University of Alberta. Since the early 1980s, his team conducted both analytical and experimental research focused on developing design procedures suitable for drafting design standards (Driver et al., 1997, Thorburn et al., 1983, Timler and Kulak, 1983, and Tromposch and Kulak, 1987). Recent research in the United States by Astaneh (2001) supports the assertion by Canadian academia that unstiffened plate, post-buckling behaviour acts as a capable shear resisting system.

Analytical Models

There are two different modelling techniques:

- Strip Model

- Modified Plate-Frame Interaction (M-PFI) model

The strip model represents shear panels as a series of inclined strip elements, capable of transmitting tension forces only, and oriented in the same direction as the average principal tensile stresses in the panel. By replacing a plate panel with struts, the resulting steel structure can be analyzed using currently available commercial computer analysis software. Research conducted at the University of British Columbia by Rezai et al. (1999) showed that the strip model is significantly incompatible and inaccurate for a wide range of SPW arrangements.

The strip model is limited mostly to SPSWs with thin plates (low critical buckling capacity) and certain ratios. In the development of this model, no solution has been provided for a perforated SPSW, shear walls with thick steel plates and shear walls with stiffeners. The strip model concept, although appropriate for practical analysis of thin plates, is not directly applicable to other types of plates. Moreover, its implementations have yet to be incorporated in commonly used commercial computer analysis software.

In order to overcome this limitation, a general method was developed for the analysis and design of SPWs within different configurations, including walls with or without openings, with thin or thick plates, and with or without stiffeners. This method considers the behavior of the steel plate and frame separately, and accounts for the interaction of these two elements, which leads to a more rational engineering design of an SPSW system. However, this model has serious shortcomings when the flexural behavior of an SPSW needs to be properly accounted for, such as the case of a slender tall building.

Modified Plate-Frame Interaction (M-PFI) model is based upon an existing shear model originally presented by Roberts and Sabouri-Ghomi (1992). Sabouri-Ghomi, Ventura and Kharrazi (2005) further refined the model and named it the Plate-Frame Interaction (PFI) model. In this paper, the PFI analytical model is then further enhanced by 'modifying' the load-displacement diagram to include the effect of overturning moments on the SPW response, hence the given name of the M-PFI model. , The method also addresses bending and shear interactions of the plastic ultimate capacity of steel panels, as well as bending and shear interactions of the ultimate yield strength for each individual component, that is the steel plate and surrounding frame.

References

- Reitherman, Robert (2012). Earthquakes and Engineers: An International History. Reston, VA: ASCE Press. pp. 356–357. ISBN 9780784410714.

- A Survey on concepts of design and executing of Superframe RC Earthquake proof Structures" (2016) by Kiarash Khodabakhshi ISBN 9783668208704.

7

Allied Fields of Earthquake Engineering

Earthquake engineering requires input from other fields to develop and design structures that can withstand earthquakes. One of these branches, known as structural engineering is concerned with the design of structures, creative use of building materials and economy in construction. Geotechnical engineering on the other hand focuses on the soil conditions and site investigations which delve into aspects like fault distribution, soil erosion, subsurface categorization, bedrock properties etc. This chapter inspects the allied fields of geotechnical engineering, civil engineering, seismic analysis and structural engineering.

Structural Engineering

Structural engineers are trained to understand and calculate the stability, strength and rigidity of built structures for buildings and nonbuilding structures, to develop designs and integrate their design with that of other designers, and to supervise construction of projects on site. They can also be involved in the design of machinery, medical equipment, vehicles etc. where structural integrity affects functioning and safety.

Structural engineering theory is based upon applied physical laws and empirical knowledge of the structural performance of different materials and geometries. Structural engineering design utilizes a number of relatively simple structural elements to build complex structural systems. Structural engineers are responsible for making creative and efficient use of funds, structural elements and materials to achieve these goals.

Structural engineering deals with the making of complex systems like the International Space Station, seen here from the departing Space Shuttle *Atlantis*.

Structural engineers investigating NASA's Mars-bound spacecraft, the Phoenix Mars Lander

Structural Engineer (Professional)

Structural engineers are responsible for engineering design and analysis. Entry-level structural engineers may design the individual structural elements of a structure, for example the beams, columns, and floors of a building. More experienced engineers may be responsible for the structural design and integrity of an entire system, such as a building.

Structural engineers often specialize in particular fields, such as bridge engineering, building engineering, pipeline engineering, industrial structures, or special mechanical structures such as vehicles, ships or aircraft.

Structural engineering has existed since humans first started to construct their own structures. It became a more defined and formalised profession with the emergence of the architecture profession as distinct from the engineering profession during the industrial revolution in the late 19th century. Until then, the architect and the structural engineer were usually one and the same - the master builder. Only with the development of specialised knowledge of structural theories that emerged during the 19th and early 20th centuries did the professional structural engineer come into existence.

The role of a structural engineer today involves a significant understanding of both static and dynamic loading, and the structures that are available to resist them. The complexity of modern structures often requires a great deal of creativity from the engineer in order to ensure the structures support and resist the loads they are subjected to. A structural engineer will typically have a four or five year undergraduate degree, followed by a minimum of three years of professional practice before being considered fully qualified. Structural engineers are licensed or accredited by different learned societies and regulatory bodies around the world (for example, the Institution of Structural Engineers in the UK). Depending on the degree course they have studied and/or the jurisdiction they are seeking licensure in, they may be accredited (or licensed) as just structural engineers, or as civil engineers, or as both civil and structural engineers. Another international organisation is IABSE (International Association for Bridge and

Structural Engineering). The aim of that association is to exchange knowledge and to advance the practice of structural engineering worldwide in the service of the profession and society.

History of Structural Engineering

Pont du Gard, France, a Roman era aqueduct circa 19 BC.

Structural engineering dates back to 2700 B.C.E. when the step pyramid for Pharaoh Djoser was built by Imhotep, the first engineer in history known by name. Pyramids were the most common major structures built by ancient civilizations because the structural form of a pyramid is inherently stable and can be almost infinitely scaled (as opposed to most other structural forms, which cannot be linearly increased in size in proportion to increased loads).

However, it's important to note that the structural stability of the pyramid is not primarily a result of its shape. The integrity of the pyramid is intact as long as each of the stones is able to support the weight of the stone above it. The limestone blocks were taken from a quarry near the build site. Since the compressive strength of limestone is anywhere from 30 to 250 MPa (MPa = Pa * 10^6), the blocks will not fail under compression. Therefore, the structural strength of the pyramid stems from the material properties of the stones from which it was built rather than the pyramid's geometry.

Throughout ancient and medieval history most architectural design and construction was carried out by artisans, such as stone masons and carpenters, rising to the role of master builder. No theory of structures existed, and understanding of how structures stood up was extremely limited, and based almost entirely on empirical evidence of 'what had worked before'. Knowledge was retained by guilds and seldom supplanted by advances. Structures were repetitive, and increases in scale were incremental.

No record exists of the first calculations of the strength of structural members or the behavior of structural material, but the profession of structural engineer only really took shape with the Industrial Revolution and the re-invention of concrete.

The physical sciences underlying structural engineering began to be understood in the Renaissance and have since developed into computer-based applications pioneered in the 1970s.

Timeline

- 1452–1519 Leonardo da Vinci made many contributions

- 1638: Galileo Galilei published the book "Two New Sciences" in which he examined the failure of simple structures

Galileo Galilei published the book "Two New Sciences" in which he examined the failure of simple structures

- 1660: Hooke's law by Robert Hooke

- 1687: Isaac Newton published "Philosophiae Naturalis Principia Mathematica" which contains the Newton's laws of motion

Isaac Newton published "Philosophiae Naturalis Principia Mathematica" which contains the Newton's laws of motion

- 1750: Euler–Bernoulli beam equation

- 1700–1782: Daniel Bernoulli introduced the principle of virtual work

- 1707–1783: Leonhard Euler developed the theory of buckling of columns

Leonhard Euler developed the theory of buckling of columns

- 1826: Claude-Louis Navier published a treatise on the elastic behaviors of structures

- 1873: Carlo Alberto Castigliano presented his dissertation "Intorno ai sistemi elastici", which contains his theorem for computing displacement as partial derivative of the strain energy. This theorem includes the method of *least work* as a special case

- 1874: Otto Mohr formalized the idea of a statically indeterminate structure.

- 1922: Timoshenko corrects the Euler-Bernoulli beam equation

- 1936: Hardy Cross' publication of the moment distribution method, an important innovation in the design of continuous frames.

- 1941: Alexander Hrennikoff solved the discretization of plane elasticity problems using a lattice framework

- 1942: R. Courant divided a domain into finite subregions

- 1956: J. Turner, R. W. Clough, H. C. Martin, and L. J. Topp's paper on the "Stiffness and Deflection of Complex Structures" introduces the name "finite-element method" and is widely recognized as the first comprehensive treatment of the method as it is known today

Structural Failure

The history of structural engineering contains many collapses and failures. Sometimes this is due to obvious negligence, as in the case of the Pétionville school collapse, in which Rev. Fortin Augustin *"constructed the building all by himself, saying he didn't need an engineer as he had good knowledge of construction"* following a partial collapse of the three-story schoolhouse that sent neighbors fleeing. The final collapse killed 94 people, mostly children.

In other cases structural failures require careful study, and the results of these inquiries have resulted in improved practices and greater understanding of the science of structural engineering. Some such studies are the result of forensic engineering investigations where the original engineer seems to have done everything in accordance with the state of the profession and acceptable practice yet a failure still eventuated. A famous case of structural knowledge and practice being advanced in this manner can be found in a series of failures involving box girders which collapsed in Australia during the 1970s.

Specializations

Building Structures

Sydney Opera House, designed by Architect Jørn Utzon and structural design by Ove Arup & Partners

Millennium Dome in London, UK, by Richard Rogers and Buro Happold

Burj Khalifa, in Dubai, the world's tallest building, shown under construction in 2007 (since completed)

Structural building engineering includes all structural engineering related to the design of buildings. It is a branch of structural engineering closely affiliated with architecture.

Structural building engineering is primarily driven by the creative manipulation of materials and forms and the underlying mathematical and scientific ideas to achieve an end which fulfills its functional requirements and is structurally safe when subjected to all the loads it could reasonably be expected to experience. This is subtly different from architectural design, which is driven by the creative manipulation of materials and forms, mass, space, volume, texture and light to achieve an end which is aesthetic, functional and often artistic.

The architect is usually the lead designer on buildings, with a structural engineer employed as a sub-consultant. The degree to which each discipline actually leads the design depends heavily on the type of structure. Many structures are structurally simple and led by architecture, such as multi-storey office buildings and housing, while other structures, such as tensile structures, shells and gridshells are heavily dependent on their form for their strength, and the engineer may have a more significant influence on the form, and hence much of the aesthetic, than the architect.

The structural design for a building must ensure that the building is able to stand up safely, able to function without excessive deflections or movements which may cause fatigue of structural elements, cracking or failure of fixtures, fittings or partitions, or discomfort for occupants. It must account for movements and forces due to temperature, creep, cracking and imposed loads. It must also ensure that the design is practically buildable within acceptable manufacturing tolerances of the materials. It must allow the architecture to work, and the building services to fit within the building and function (air conditioning, ventilation, smoke extract, electrics, lighting etc.). The structural design of a modern building can be extremely complex, and often requires a large team to complete.

Structural engineering specialties for buildings include:

- Earthquake engineering
- Façade engineering
- Fire engineering
- Roof engineering
- Tower engineering
- Wind engineering

Earthquake Engineering Structures

Earthquake engineering structures are those engineered to withstand earthquakes.

Earthquake-proof pyramid El Castillo, Chichen Itza

The main objectives of earthquake engineering are to understand the interaction of structures with the shaking ground, foresee the consequences of possible earthquakes, and design and construct the structures to perform during an earthquake.

Earthquake-proof structures are not necessarily extremely strong like the El Castillo pyramid at Chichen Itza shown above. In fact, many structures considered strong may in fact be stiff, which can result in poor seismic performance.

One important tool of earthquake engineering is base isolation, which allows the base of a structure to move freely with the ground.

Civil Engineering Structures

Civil structural engineering includes all structural engineering related to the built environment. It includes:

• Bridges	• Power stations
• Dams	• Railways
• Earthworks	• Retaining structures and walls
• Foundations	• Roads
• Offshore structures	• Tunnels
• Pipelines	• Waterways
	• reservoir
	• Water and wastewater infrastructure

The structural engineer is the lead designer on these structures, and often the sole designer. In the design of structures such as these, structural safety is of paramount importance (in the UK, designs for dams, nuclear power stations and bridges must be signed off by a chartered engineer).

Civil engineering structures are often subjected to very extreme forces, such as large variations in temperature, dynamic loads such as waves or traffic, or high pressures from water or compressed gases. They are also often constructed in corrosive environments, such as at sea, in industrial facilities or below ground.

Mechanical Structures

Car

Motorbike

Principles of structural engineering are applied to variety of mechanical (moveable) structures. The design of static structures assumes they always have the same geometry (in fact, so-called static structures can move significantly, and structural engineering design must take this into account where necessary), but the design of moveable or moving structures must account for fatigue, variation in the method in which load is resisted and significant deflections of structures.

The forces which parts of a machine are subjected to can vary significantly, and can do so at a great rate. The forces which a boat or aircraft are subjected to vary enormously and will do so thousands of times over the structure's lifetime. The structural design must ensure that such structures are able to endure such loading for their entire design life without failing.

These works can require mechanical structural engineering:

- Boilers and pressure vessels
- Coachworks and carriages
- Cranes
- Elevators

- Escalators

- Marine vessels and hulls

Aerospace Structures

An Airbus A380, the world's largest passenger airliner

Design of missile needs in depth understanding of Structural Analysis

Aerospace structure types include launch vehicles, (Atlas, Delta, Titan), missiles (ALCM, Harpoon), Hypersonic vehicles (Space Shuttle), military aircraft (F-16, F-18) and commercial aircraft (Boeing 777, MD-11). Aerospace structures typically consist of thin plates with stiffeners for the external surfaces, bulkheads and frames to support the shape and fasteners such as welds, rivets, screws and bolts to hold the components together.

Nanoscale Structures

A nanostructure is an object of intermediate size between molecular and microscopic (micrometer-sized) structures. In describing nanostructures it is necessary to differentiate between the number of dimensions on the nanoscale. Nanotextured surfaces have one dimension on the nanoscale, i.e., only the thickness of the surface of an object is between 0.1 and 100 nm. Nanotubes have two dimensions on the nanoscale, i.e., the diameter of the tube is between 0.1 and 100 nm; its length could be much greater. Finally, spherical nanoparticles have three dimensions on the nanoscale, i.e., the particle is between 0.1 and 100 nm in each spatial dimension. The terms nanoparticles and ultrafine particles (UFP) often are used synonymously although UFP can reach into the micrometre range. The term 'nanostructure' is often used when referring to magnetic technology.

Structural Engineering for Medical Science

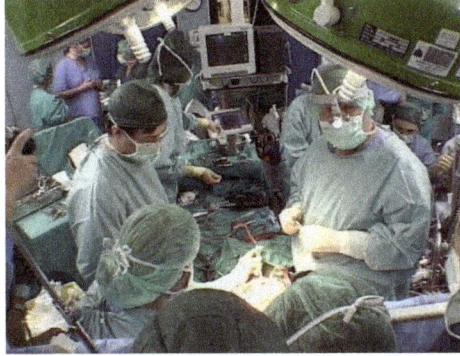

Designing Medical Equipment needs in-depth understanding of Structural Engineering

Medical equipment (also known as armamentarium) is designed to aid in the diagnosis, monitoring or treatment of medical conditions. There are several basic types: Diagnostic equipment includes medical imaging machines, used to aid in diagnosis ; equipment includes infusion pumps, medical lasers and LASIK surgical machines ; Medical monitors allow medical staff to measure a patient›s medical state. Monitors may measure patient vital signs and other parameters including ECG, EEG, blood pressure, and dissolved gases in the blood ; Diagnostic Medical Equipment may also be used in the home for certain purposes, e.g. for the control of diabetes mellitus. A biomedical equipment technician (BMET) is a vital component of the healthcare delivery system. Employed primarily by hospitals, BMETs are the people responsible for maintaining a facility's medical equipment.

Structural Elements

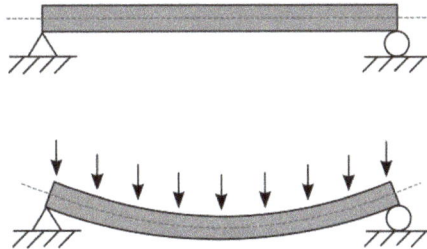

A statically determinate simply supported beam, bending under an evenly distributed load.

Any structure is essentially made up of only a small number of different types of elements:

- Columns

- Beams

- Plates

- Arches

- Shells

- Catenaries

Many of these elements can be classified according to form (straight, plane / curve) and dimensionality (one-dimensional / two-dimensional):

	One-dimensional		Two-dimensional	
	straight	curve	plane	curve
(predominantly) bending	beam	continuous arch	plate, concrete slab	lamina, dome
(predominant) tensile stress	rope, tie	Catenary	shell	
(predominant) compression	pier, column		Load-bearing wall	

Columns

Columns are elements that carry only axial force - compression - or both axial force and bending (which is technically called a beam-column but practically, just a column). The design of a column must check the axial capacity of the element, and the buckling capacity.

The buckling capacity is the capacity of the element to withstand the propensity to buckle. Its capacity depends upon its geometry, material, and the effective length of the column, which depends upon the restraint conditions at the top and bottom of the column. The effective length is where is the real length of the column and K is the factor dependent on the restraint conditions.

The capacity of a column to carry axial load depends on the degree of bending it is subjected to, and vice versa. This is represented on an interaction chart and is a complex non-linear relationship.

Beams

A beam may be defined as an element in which one dimension is much greater than the other two and the applied loads are usually normal to the main axis of the element. Beams and columns are called line elements and are often represented by simple lines in structural modeling.

Little Belt: a truss bridge in Denmark

- cantilevered (supported at one end only with a fixed connection)

- simply supported (supported vertically at each end; horizontally on only one to withstand friction, and able to rotate at the supports)

- fixed (supported at both ends by fixed connection; unable to rotate at the supports)

- continuous (supported by three or more supports)

- a combination of the above (ex. supported at one end and in the middle)

Beams are elements which carry pure bending only. Bending causes one part of the section of a beam (divided along its length) to go into compression and the other part into tension. The compression part must be designed to resist buckling and crushing, while the tension part must be able to adequately resist the tension.

Trusses

The McDonnell Planetarium by Gyo Obata in St Louis, Missouri, USA, a concrete shell structure

A truss is a structure comprising two types of structural elements; compression members and tension members (i.e. struts and ties). Most trusses use gusset plates to connect intersecting elements. Gusset plates are relatively flexible and minimize bending moments at the connections, thus allowing the truss members to carry primarily tension or compression.

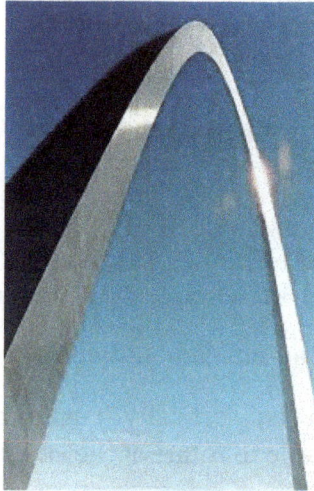

The 630 foot (192 m) high, stainless-clad (type 304)
Gateway Arch in Saint Louis, Missouri

Trusses are usually utilised in large-span structures, where it would be uneconomical to use solid beams.

Plates

Plates carry bending in two directions. A concrete flat slab is an example of a plate. Plates are understood by using continuum mechanics, but due to the complexity involved they are most often designed using a codified empirical approach, or computer analysis.

They can also be designed with yield line theory, where an assumed collapse mechanism is analysed to give an upper bound on the collapse load. This technique is used in practice but because the method provides an upper-bound, i.e. an unsafe prediction of the collapse load, for poorly conceived collapse mechanisms great care is needed to ensure that the assumed collapse mechanism is realistic.

Shells

Shells derive their strength from their form, and carry forces in compression in two directions. A dome is an example of a shell. They can be designed by making a hanging-chain model, which will act as a catenary in pure tension, and inverting the form to achieve pure compression.

Arches

Arches carry forces in compression in one direction only, which is why it is appropriate to build arches out of masonry. They are designed by ensuring that the line of thrust of the force remains within the depth of the arch. It is mainly used to increase the bountifulness of any structure.

Catenaries

Catenaries derive their strength from their form, and carry transverse forces in pure tension by deflecting (just as a tightrope will sag when someone walks on it). They are almost always cable or fabric structures. A fabric structure acts as a catenary in two directions.

Structural Engineering Theory

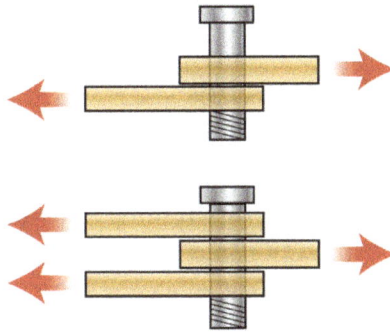

Figure of a bolt in shear stress. Top figure illustrates single shear, bottom figure illustrates double shear.

Structural engineering depends upon a detailed knowledge of applied mechanics, materials science and applied mathematics to understand and predict how structures support and resist self-weight and imposed loads. To apply the knowledge successfully a structural engineer generally requires detailed knowledge of relevant empirical and theoretical design codes, the techniques of structural analysis, as well as some knowledge of the corrosion resistance of the materials and structures, especially when those structures are exposed to the external environment. Since the 1990s, specialist software has become available to aid in the design of structures, with the functionality to assist in the drawing, analyzing and designing of structures with maximum precision; examples include AutoCAD, StaadPro, ETABS, Prokon, Revit Structure etc. Such software may also take into consideration environmental loads, such as from earthquakes and winds.

Materials

Structural engineering depends on the knowledge of materials and their properties, in order to understand how different materials support and resist loads.

Common structural materials are:

- Iron: Wrought iron, Cast iron

- Concrete: Reinforced concrete, Prestressed concrete

- Alloy: Steel, Stainless steel

- Masonry

- Timber: Hardwood, Softwood

- Aluminium

- Composite materials: Plywood

- Other structural materials:Adobe, Bamboo, Carbon fibre, Fiber reinforced plastic, Mudbrick, Roofing materials

Example of Structural Engineering

Structural Dynamics

Structural analysis is mainly concerned with finding out the behavior of a physical structure when subjected to force. This action can be in the form of load due to the weight of things such as people, furniture, wind, snow, etc. or some other kind of excitation such as an earthquake, shaking of the ground due to a blast nearby, etc. In essence all these loads are dynamic, including the self-weight of the structure because at some point in time these loads were not there. The distinction is made between the dynamic and the static analysis on the basis of whether the applied action has enough acceleration in comparison to the structure's natural frequency. If a load is applied sufficiently slowly, the inertia forces (Newton's first law of motion) can be ignored and the analysis can be simplified as static analysis. Structural dynamics, therefore, is a type of structural analysis which covers the behavior of structures subjected to dynamic (actions having high acceleration) loading. Dynamic loads include people, wind, waves, traffic, earthquakes, and blasts. Any structure can be subjected to dynamic loading. Dynamic analysis can be used to find dynamic displacements, time history, and modal analysis.

A dynamic analysis is also related to the inertia forces developed by a structure when it is excited by means of dynamic loads applied suddenly (e.g., wind blasts, explosion, earthquake).

A static load is one which varies very slowly. A dynamic load is one which changes with time fairly quickly in comparison to the structure's natural frequency. If it changes slowly, the structure's response may be determined with static analysis, but if it varies quickly (relative to the structure's ability to respond), the response must be determined with a dynamic analysis.

Dynamic analysis for simple structures can be carried out manually, but for complex

structures finite element analysis can be used to calculate the mode shapes and frequencies.

Displacements

A dynamic load can have a significantly larger effect than a static load of the same mag-
nitude due to the structure's inability to respond quickly to the loading (by deflecting).
The increase in the effect of a dynamic load is given by the dynamic amplification factor
(DAF):

$$DAF = \frac{u_{max}}{u_{static}}$$

where u is the deflection of the structure due to the applied load.

Graphs of dynamic amplification factors vs non-dimensional rise time (t_r/T) exist for
standard loading functions. Hence the DAF for a given loading can be read from the
graph, the static deflection can be easily calculated for simple structures and the
dynamic deflection found.

Time History Analysis

A full time history will give the response of a structure over time during and after the
application of a load. To find the full time history of a structure's response, you must
solve the structure's equation of motion.

Example

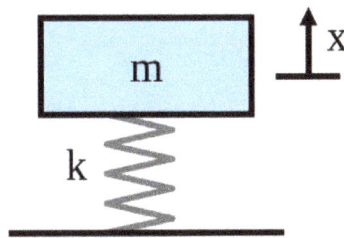

A simple single degree of freedom system (a mass, M, on a spring of stiffness k, for example)
has the following equation of motion:

$$M\ddot{x} + kx = F(t)$$

If the loading F(t) is a Heaviside step function (the sudden application of a constant
load), the solution to the equation of motion is:

$$x = \frac{F_0}{k}[1 - cos(\omega t)]$$

Where $\omega = \sqrt{\dfrac{k}{M}}$ and the fundamental natural frequency, $f = \dfrac{\omega}{2\pi}$.

The static deflection of a single degree of freedom system is:

$$x_{static} = \frac{F_0}{k}$$

so you can write, by combining the above formulae:

$$x = x_{static}[1 - cos(\omega t)]$$

This gives the (theoretical) time history of the structure due to a load F(t), where the false assumption is made that there is no damping.

Although this is too simplistic to apply to a real structure, the Heaviside Step Function is a reasonable model for the application of many real loads, such as the sudden addition of a piece of furniture, or the removal of a prop to a newly cast concrete floor. However, in reality loads are never applied instantaneously - they build up over a period of time (this may be very short indeed). This time is called the rise time.

As the number of degrees of freedom of a structure increases it very quickly becomes too difficult to calculate the time history manually - real structures are analysed using non-linear finite element analysis software.

Damping

Any real structure will dissipate energy (mainly through friction). This can be modelled by modifying the DAF

$$DAF = 1 + e^{-c\pi}$$

Where $c = \dfrac{\text{Damping Coefficient}}{\text{Critical Damping Coefficient}}$ and is typically 2%-10% depending on the type of construction:

- Bolted steel ~6%

- Reinforced concrete ~ 5%

- Welded steel ~ 2%

- Brick masonry ~ 10%

Generally damping would be ignored for non-transient events (such as wind loading or crowd loading), but would be important for transient events (for example, an impulse load such as an earthquake loading or bomb blast).

Modal Analysis

A modal analysis calculates the frequency modes or natural frequencies of a given system, but not necessarily its full-time history response to a given input. The natural frequency of a system is dependent only on the stiffness of the structure and the mass which participates with the structure (including self-weight). It is not dependent on the load function.

It is useful to know the modal frequencies of a structure as it allows you to ensure that the frequency of any applied periodic loading will not coincide with a modal frequency and hence cause resonance, which leads to large oscillations.

The method is:

1. Find the natural modes (the shape adopted by a structure) and natural frequencies

2. Calculate the response of each mode

3. Optionally superpose the response of each mode to find the full modal response to a given loading

Energy Method

It is possible to calculate the frequency of different mode shape of system manually by the energy method. For a given mode shape of a multiple degree of freedom system you can find an "equivalent" mass, stiffness and applied force for a single degree of freedom system. For simple structures the basic mode shapes can be found by inspection, but it is not a conservative method. Rayleigh's principle states:

"The frequency ω of an arbitrary mode of vibration, calculated by the energy method, is always greater than - or equal to - the fundamental frequency ω_n."

For an assumed mode shape $\bar{u}(x)$ of a structural system with mass M; bending stiffness, EI (Young's modulus, E, multiplied by the second moment of area, I); and applied force, F(x):

$$\text{Equivalent mass, } M_{eq} = \int M\bar{u}^2 du$$

$$\text{Equivalent stiffness, } k_{eq} = \int EI(\frac{d^2\bar{u}}{dx^2})^2 dx$$

$$\text{Equivalent force, } F_{eq} = \int F\bar{u}dx$$

then, as above:

$$\omega = \sqrt{\frac{k_{eq}}{M_{eq}}}$$

Modal Response

The complete modal response to a given load F(x,t) is $v(x,t) = \sum u_n(x,t)$. he summation can be carried out by one of three common methods:

- Superpose complete time histories of each mode (time consuming, but exact)

- Superpose the maximum amplitudes of each mode (quick but conservative)

- Superpose the square root of the sum of squares (good estimate for well-separated frequencies, but unsafe for closely spaced frequencies)

To superpose the individual modal responses manually, having calculated them by the energy method:

Assuming that the rise time t_r is known (T = 2π/ω), it is possible to read the DAF from a standard graph. The static displacement can be calculated with $u_{static} = \dfrac{F_{1,eq}}{k_{1,eq}}$. The dynamic displacement for the chosen mode and applied force can then be found from:

$$u_{max} = u_{static} DAF$$

Modal Participation Factor

For real systems there is often mass participating in the forcing function (such as the mass of ground in an earthquake) and mass participating in inertia effects (the mass of the structure itself, M_{eq}). The modal participation factor Γ is a comparison of these two masses. For a single degree of freedom system Γ = 1.

$$\Gamma = \frac{\sum M_n \bar{u}_n}{\sum M_n \bar{u}_n^2}$$

Geotechnical Engineering

Geotechnical engineering is the branch of civil engineering concerned with the engineering behavior of earth materials. Geotechnical engineering is important in civil engineering, but also has applications in military, mining, petroleum and other engineering disciplines that are concerned with construction occurring on the surface or within the ground. Geotechnical engineering uses principles of soil mechanics and rock mechanics to investigate subsurface conditions and materials; determine the relevant physical/mechanical and chemical properties of these materials; evaluate stability of natural slopes and man-made soil deposits; assess risks posed by site conditions; design earthworks and structure foundations; and monitor site conditions, earthwork and foundation construction.

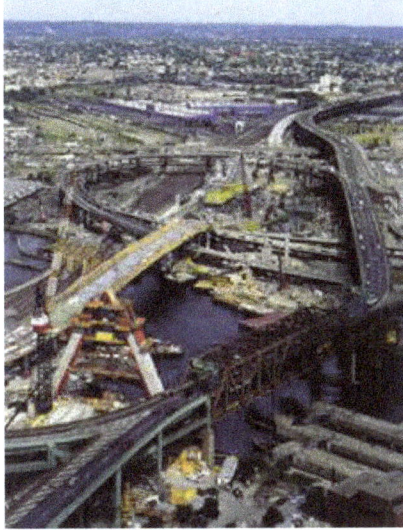

Boston's Big Dig presented geotechnical challenges in an urban environment.

A typical geotechnical engineering project begins with a review of project needs to define the required material properties. Then follows a site investigation of soil, rock, fault distribution and bedrock properties on and below an area of interest to determine their engineering properties including how they will interact with, on or in a proposed construction. Site investigations are needed to gain an understanding of the area in or on which the engineering will take place. Investigations can include the assessment of the risk to humans, property and the environment from natural hazards such as earthquakes, landslides, sinkholes, soil liquefaction, debris flows and rockfalls.

A geotechnical engineer then determines and designs the type of foundations, earthworks, and/or pavement subgrades required for the intended man-made structures to be built. Foundations are designed and constructed for structures of various sizes such as high-rise buildings, bridges, medium to large commercial buildings, and smaller structures where the soil conditions do not allow code-based design.

Foundations built for above-ground structures include shallow and deep foundations. Retaining structures include earth-filled dams and retaining walls. Earthworks include embankments, tunnels, dikes and levees, channels, reservoirs, deposition of hazardous waste and sanitary landfills.

Geotechnical engineering is also related to coastal and ocean engineering. Coastal engineering can involve the design and construction of wharves, marinas, and jetties. Ocean engineering can involve foundation and anchor systems for offshore structures such as oil platforms.

The fields of geotechnical engineering and engineering geology are closely related, and have large areas of overlap. However, the field of geotechnical engineering is a specialty of engineering, where the field of engineering geology is a specialty of geology.

History

Humans have historically used soil as a material for flood control, irrigation purposes, burial sites, building foundations, and as construction material for buildings. First activities were linked to irrigation and flood control, as demonstrated by traces of dykes, dams, and canals dating back to at least 2000 BCE that were found in ancient Egypt, ancient Mesopotamia and the Fertile Crescent, as well as around the early settlements of Mohenjo Daro and Harappa in the Indus valley. As the cities expanded, structures were erected supported by formalized foundations; Ancient Greeks notably constructed pad footings and strip-and-raft foundations. Until the 18th century, however, no theoretical basis for soil design had been developed and the discipline was more of an art than a science, relying on past experience.

Several foundation-related engineering problems, such as the Leaning Tower of Pisa, prompted scientists to begin taking a more scientific-based approach to examining the subsurface. The earliest advances occurred in the development of earth pressure theories for the construction of retaining walls. Henri Gautier, a French Royal Engineer, recognized the "natural slope" of different soils in 1717, an idea later known as the soil's angle of repose. A rudimentary soil classification system was also developed based on a material's unit weight, which is no longer considered a good indication of soil type.

The application of the principles of mechanics to soils was documented as early as 1773 when Charles Coulomb (a physicist, engineer, and army Captain) developed improved methods to determine the earth pressures against military ramparts. Coulomb observed that, at failure, a distinct slip plane would form behind a sliding retaining wall and he suggested that the maximum shear stress on the slip plane, for design purposes, was the sum of the soil cohesion, , and friction , where is the normal stress on the slip plane and is the friction angle of the soil. By combining Coulomb's theory with Christian Otto Mohr's 2D stress state, the theory became known as Mohr-Coulomb theory. Although it is now recognized that precise determination of cohesion is impossible because is not a fundamental soil property, the Mohr-Coulomb theory is still used in practice today.

In the 19th century Henry Darcy developed what is now known as Darcy's Law describing the flow of fluids in porous media. Joseph Boussinesq (a mathematician and physicist) developed theories of stress distribution in elastic solids that proved useful for estimating stresses at depth in the ground; William Rankine, an engineer and physicist, developed an alternative to Coulomb's earth pressure theory. Albert Atterberg developed the clay consistency indices that are still used today for soil classification. Osborne Reynolds recognized in 1885 that shearing causes volumetric dilation of dense and contraction of loose granular materials.

Modern geotechnical engineering is said to have begun in 1925 with the publication of *Erdbaumechanik* by Karl Terzaghi (a civil engineer and geologist). Considered by many to be the father of modern soil mechanics and geotechnical engineering, Terzaghi developed the principle of effective stress, and demonstrated that the shear strength of soil is controlled by effective stress. Terzaghi also developed the framework for theories of bearing capacity of foundations, and the theory for prediction of the rate of settlement of clay layers due to consolidation. In his 1948 book, Donald Taylor recognized that interlocking and dilation of densely packed particles contributed to the peak strength of a soil. The interrelationships between volume change behavior (dilation, contraction, and consolidation) and shearing behavior were all connected via the theory of plasticity using critical state soil mechanics by Roscoe, Schofield, and Wroth with the publication of "On the Yielding of Soils" in 1958. Critical state soil mechanics is the basis for many contemporary advanced constitutive models describing the behavior of soil.

Geotechnical centrifuge modeling is a method of testing physical scale models of geotechnical problems. The use of a centrifuge enhances the similarity of the scale model tests involving soil because the strength and stiffness of soil is very sensitive to the confining pressure. The centrifugal acceleration allows a researcher to obtain large (prototype-scale) stresses in small physical models.

Practicing Engineers

Geotechnical engineers are typically graduates of a four-year civil engineering program and some hold a masters degree. In the USA, geotechnical engineers are typically licensed and regulated as Professional Engineers (PEs) in most states; currently only California and Oregon have licensed geotechnical engineering specialties. The Academy of Geo-Professionals (AGP) began issuing Diplomate, Geotechnical Engineering (D.GE) certification in 2008. State governments will typically license engineers who have graduated from an ABET accredited school, passed the Fundamentals of Engineering examination, completed several years of work experience under the supervision of a licensed Professional Engineer, and passed the Professional Engineering examination.

Soil Mechanics

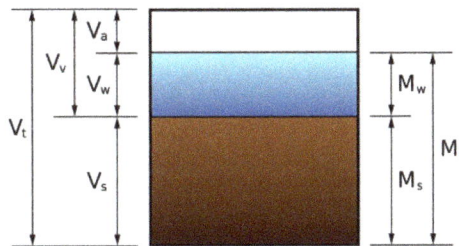

A phase diagram of soil indicating the weights and volumes of air, soil, water, and voids.

In geotechnical engineering, soils are considered a three-phase material composed of: rock or mineral particles, water and air. The voids of a soil, the spaces in between mineral particles, contain the water and air.

The engineering properties of soils are affected by four main factors: the predominant size of the mineral particles, the type of mineral particles, the grain size distribution, and the relative quantities of mineral, water and air present in the soil matrix. Fine particles (fines) are defined as particles less than 0.075 mm in diameter.

Soil Properties

Some of the important properties of soils that are used by geotechnical engineers to analyze site conditions and design earthworks, retaining structures, and foundations are:

Specific Weight or Unit Weight

Cumulative weight of the solid particles, water and air of the unit volume of soil. Note that the air phase is often assumed to be weightless.

Porosity

Ratio of the volume of voids (containing air, water, or other fluids) in a soil to the total volume of the soil. Porosity is mathematically related to void ratio the by

$$n = \frac{e}{1+e}$$

here e is void ratio and n is porosity

Void Ratio

The ratio of the volume of voids to the volume of solid particles in a soil mass. Void ratio is mathematically related to the porosity by

$$e = \frac{n}{1-n}$$

Permeability

A measure of the ability of water to flow through the soil. It is expressed in units of velocity.

Compressibility

The rate of change of volume with effective stress. If the pores are filled with water, then the water must be squeezed out of the pores to allow volumetric compression of the soil; this process is called consolidation.

Shear Strength

The maximum shear stress that can be applied in a soil mass without causing shear failure.

Atterberg Limits

Liquid limit, Plastic limit, and Shrinkage limit. These indices are used for estimation of other engineering properties and for soil classification.

Geotechnical Investigation

Geotechnical engineers and engineering geologists perform geotechnical investigations to obtain information on the physical properties of soil and rock underlying (and sometimes adjacent to) a site to design earthworks and foundations for proposed structures, and for repair of distress to earthworks and structures caused by subsurface conditions. A geotechnical investigation will include surface exploration and subsurface exploration of a site. Sometimes, geophysical methods are used to obtain data about sites. Subsurface exploration usually involves in-situ testing (two common examples of in-situ tests are the standard penetration test and cone penetration test). In addition site investigation will often include subsurface sampling and laboratory testing of the soil samples retrieved. The digging of test pits and trenching (particularly for locating faults and slide planes) may also be used to learn about soil conditions at depth. Large diameter borings are rarely used due to safety concerns and expense, but are sometimes used to allow a geologist or engineer to be lowered into the borehole for direct visual and manual examination of the soil and rock stratigraphy.

A variety of soil samplers exist to meet the needs of different engineering projects. The standard penetration test (SPT), which uses a thick-walled split spoon sampler, is the most common way to collect disturbed samples. Piston samplers, employing a thin-walled tube, are most commonly used for the collection of less disturbed samples. More advanced methods, such as ground freezing and the Sherbrooke block sampler, are superior, but even more expensive.

Atterberg limits tests, water content measurements, and grain size analysis, for example, may be performed on disturbed samples obtained from thick walled soil samplers. Properties such as shear strength, stiffness hydraulic conductivity, and coefficient of consolidation may be significantly altered by sample disturbance. To measure these properties in the laboratory, high quality sampling is required. Common tests to measure the strength and stiffness include the triaxial shear and unconfined compression test.

Surface exploration can include geologic mapping, geophysical methods, and photogrammetry; or it can be as simple as an engineer walking around to observe the physical conditions at the site. Geologic mapping and interpretation of geomorphology is typically completed in consultation with a geologist or engineering geologist.

Geophysical exploration is also sometimes used. Geophysical techniques used for subsurface exploration include measurement of seismic waves (pressure, shear, and Rayleigh waves), surface-wave methods and/or downhole methods, and electromagnetic surveys (magnetometer, resistivity, and ground-penetrating radar).

Foundations

A building's foundation transmits loads from buildings and other structures to the earth. Geotechnical engineers design foundations based on the load characteristics of the structure and the properties of the soils and/or bedrock at the site. In general, geotechnical engineers:

1. Estimate the magnitude and location of the loads to be supported.

2. Develop an investigation plan to explore the subsurface.

3. Determine necessary soil parameters through field and lab testing (e.g., consolidation test, triaxial shear test, vane shear test, standard penetration test).

4. Design the foundation in the safest and most economical manner.

The primary considerations for foundation support are bearing capacity, settlement, and ground movement beneath the foundations. Bearing capacity is the ability of the site soils to support the loads imposed by buildings or structures. Settlement occurs under all foundations in all soil conditions, though lightly loaded structures or rock sites may experience negligible settlements. For heavier structures or softer sites, both overall settlement relative to unbuilt areas or neighboring buildings, and differential settlement under a single structure, can be concerns. Of particular concern is settlement which occurs over time, as immediate settlement can usually be compensated for during construction. Ground movement beneath a structure's foundations can occur due to shrinkage or swell of expansive soils due to climatic changes, frost expansion of soil, melting of permafrost, slope instability, or other causes. All these factors must be considered during design of foundations.

Many building codes specify basic foundation design parameters for simple conditions, frequently varying by jurisdiction, but such design techniques are normally limited to certain types of construction and certain types of sites, and are frequently very conservative.

In areas of shallow bedrock, most foundations may bear directly on bedrock; in other areas, the soil may provide sufficient strength for the support of structures. In areas of deeper bedrock with soft overlying soils, deep foundations are used to support structures directly on the bedrock; in areas where bedrock is not economically available, stiff "bearing layers" are used to support deep foundations instead.

Shallow Foundations

Shallow foundations are a type of foundation that transfers building load to the very near the surface, rather than to a subsurface layer. Shallow foundations typically have a depth to width ratio of less than 1.

Example of a slab-on-grade foundation.

Footings

Footings (often called "spread footings" because they spread the load) are structural elements which transfer structure loads to the ground by direct areal contact. Footings can be isolated footings for point or column loads, or strip footings for wall or other long (line) loads. Footings are normally constructed from reinforced concrete cast directly onto the soil, and are typically embedded into the ground to penetrate through the zone of frost movement and/or to obtain additional bearing capacity.

Slab Foundations

A variant on spread footings is to have the entire structure bear on a single slab of concrete underlying the entire area of the structure. Slabs must be thick enough to provide sufficient rigidity to spread the bearing loads somewhat uniformly, and to minimize differential settlement across the foundation. In some cases, flexure is allowed and the building is constructed to tolerate small movements of the foundation instead. For small structures, like single-family houses, the slab may be less than 300 mm thick; for larger structures, the foundation slab may be several meters thick.

Slab foundations can be either slab-on-grade foundations or embedded foundations, typically in buildings with basements. Slab-on-grade foundations must be designed to allow for potential ground movement due to changing soil conditions.

Deep Foundations

Deep foundations are used for structures or heavy loads when shallow foundations cannot provide adequate capacity, due to size and structural limitations. They may also be used to transfer building loads past weak or compressible soil layers. While shallow foundations rely solely on the bearing capacity of the soil beneath them, deep foundations can rely on end bearing resistance, frictional resistance along their length, or both in developing the required capacity. Geotechnical engineers use specialized tools, such as the cone penetration test, to estimate the amount of skin and end bearing resistance available in the subsurface.

Pile-driving for a bridge in Napa, California.

There are many types of deep foundations including piles, drilled shafts, caissons, piers, and earth stabilized columns. Large buildings such as skyscrapers typically require deep foundations. For example, the Jin Mao Tower in China uses tubular steel piles about 1m (3.3 feet) driven to a depth of 83.5m (274 feet) to support its weight.

In buildings that are constructed and found to undergo settlement, underpinning piles can be used to stabilise the existing building.

There are three ways to place piles for a deep foundation. They can be driven, drilled, or installed by use of an auger. Driven piles are extended to their necessary depths with the application of external energy in the same way a nail is hammered. There are four typical hammers used to drive such piles: drop hammers, diesel hammers, hydraulic hammers, and air hammers. Drop hammers simply drop a heavy weight onto the pile to drive it, while diesel hammers use a single cylinder diesel engine to force piles through the Earth. Similarly, hydraulic and air hammers supply energy to piles through hydraulic and air forces. Energy imparted from a hammer head varies with type of hammer chosen, and can be as high as a million foot pounds for large scale diesel hammers, a very common hammer head used in practice. Piles are made of a variety of material including steel, timber, and concrete. Drilled piles are created by first drilling a hole to the appropriate depth, and filling it with concrete. Drilled piles can typically carry more load than driven piles, simply due to a larger diameter pile. The auger method of pile installation is similar to drilled pile installation, but concrete is pumped into the hole as the auger is being removed.

Lateral Earth Support Structures

A retaining wall is a structure that holds back earth. Retaining walls stabilize soil and rock from downslope movement or erosion and provide support for vertical or near-vertical grade changes. Cofferdams and bulkheads, structures to hold back water, are sometimes also considered retaining walls.

The primary geotechnical concern in design and installation of retaining walls is that the weight of the retained material is creates lateral earth pressure behind the wall, which can cause the wall to deform or fail. The lateral earth pressure depends on the height of the wall, the density of the soil,the strength of the soil, and the amount of allowable movement of the wall. This pressure is smallest at the top and increases toward the bottom in a manner similar to hydraulic pressure, and tends to push the wall away from the backfill. Groundwater behind the wall that is not dissipated by a drainage system causes an additional horizontal hydraulic pressure on the wall.

Gravity Walls

Gravity walls depend on the size and weight of the wall mass to resist pressures from behind. Gravity walls will often have a slight setback, or batter, to improve wall stability. For short, landscaping walls, gravity walls made from dry-stacked (mortarless) stone or segmental concrete units (masonry units) are commonly used.

Earlier in the 20th century, taller retaining walls were often gravity walls made from large masses of concrete or stone. Today, taller retaining walls are increasingly built as composite gravity walls such as: geosynthetic or steel-reinforced backfill soil with precast facing; gabions (stacked steel wire baskets filled with rocks), crib walls (cells built up log cabin style from precast concrete or timber and filled with soil or free draining gravel) or soil-nailed walls (soil reinforced in place with steel and concrete rods).

For reinforced-soil gravity walls, the soil reinforcement is placed in horizontal layers throughout the height of the wall. Commonly, the soil reinforcement is geogrid, a high-strength polymer mesh, that provide tensile strength to hold soil together. The wall face is often of precast, segmental concrete units that can tolerate some differential movement. The reinforced soil's mass, along with the facing, becomes the gravity wall. The reinforced mass must be built large enough to retain the pressures from the soil behind it. Gravity walls usually must be a minimum of 30 to 40 percent as deep (thick) as the height of the wall, and may have to be larger if there is a slope or surcharge on the wall.

Cantilever Walls

Prior to the introduction of modern reinforced-soil gravity walls, cantilevered walls were the most common type of taller retaining wall. Cantilevered walls are made from a relatively thin stem of steel-reinforced, cast-in-place concrete or mortared masonry (often in the shape of an inverted T). These walls cantilever loads (like a beam) to a large, structural footing; converting horizontal pressures from behind the wall to vertical pressures on the ground below. Sometimes cantilevered walls are buttressed on the front, or include a counterfort on the back, to improve their stability against high loads. Buttresses are short wing walls at right angles to the main trend of the wall. These walls require rigid concrete footings below seasonal frost depth. This type of wall uses much less material than a traditional gravity wall.

Cantilever walls resist lateral pressures by friction at the base of the wall and/or passive earth pressure, the tendency of the soil to resist lateral movement.

Basements are a form of cantilever walls, but the forces on the basement walls are greater than on conventional walls because the basement wall is not free to move.

Excavation Shoring

Shoring of temporary excavations frequently requires a wall design which does not extend laterally beyond the wall, so shoring extends below the planned base of the excavation. Common methods of shoring are the use of sheet piles or soldier beams and lagging. Sheet piles are a form of driven piling using thin interlocking sheets of steel to obtain a continuous barrier in the ground, and are driven prior to excavation. Soldier beams are constructed of wide flange steel H sections spaced about 2–3 m apart, driven prior to excavation. As the excavation proceeds, horizontal timber or steel sheeting (lagging) is inserted behind the H pile flanges.

In some cases, the lateral support which can be provided by the shoring wall alone is insufficient to resist the planned lateral loads; in this case additional support is provided by walers or tie-backs. Walers are structural elements which connect across the excavation so that the loads from the soil on either side of the excavation are used to resist each other, or which transfer horizontal loads from the shoring wall to the base of the excavation. Tie-backs are steel tendons drilled into the face of the wall which extend beyond the soil which is applying pressure to the wall, to provide additional lateral resistance to the wall.

Earthworks

Excavation

Excavation is the process of training earth according to requirement by removing the soil from the site.

Filling

Filling is the process of training earth according to requirement by placing the soil on the site.

Compaction

Compaction is the process by which the density of soil is increased and permeability of soil is decreased. Fill placement work often has specifications requiring a specific degree of compaction, or alternatively, specific properties of the compacted soil. In-situ soils can be compacted by rolling, deep dynamic compaction, vibration, blasting, gyrating, kneading, compaction grouting etc.

A compactor/roller operated by U.S. Navy Seabees

Ground Improvement

Ground Improvement is a technique that improves the engineering properties of the treated soil mass. Usually, the properties modified are shear strength, stiffness and permeability. Ground improvement has developed into a sophisticated tool to support foundations for a wide variety of structures. Properly applied, i.e. after giving due consideration to the nature of the ground being improved and the type and sensitivity of the structures being built, ground improvement often reduces direct costs and saves time.

Slope Stabilization

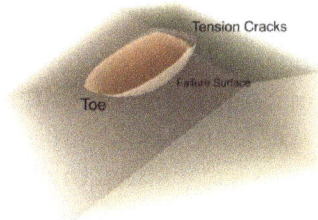

Simple slope slip section.

Slope stability is the potential of soil covered slopes to withstand and undergo movement. Stability is determined by the balance of shear stress and shear strength. A previously stable slope may be initially affected by preparatory factors, making the slope conditionally unstable. Triggering factors of a slope failure can be climatic events can then make a slope actively unstable, leading to mass movements. Mass movements can be caused by increases in shear stress, such as loading, lateral pressure, and transient forces. Alternatively, shear strength may be decreased by weathering, changes in pore water pressure, and organic material.

Several modes of failure for earth slopes include falls, topples, slides, and flows. In slopes with coarse grained soil or rocks, falls typically occur as the rapid descent of rocks and other loose slope material. A slope topples when a large column of soil tilts over its vertical axis at failure. Typical slope stability analysis considers sliding failures, categorized mainly as rotational slides or translational slides. As implied by the name, rotational slides fail along a generally curved surface, while translational slides fail along a more planar surface. A slope failing as a flow would resemble a fluid flowing downhill.

Slope Stability Analysis

Stability analysis is needed for the design of engineered slopes and for estimating the risk of slope failure in natural or designed slopes. A common assumption is that a slope consists of a layer of soil sitting on top of a rigid base. The mass and the base are assumed to interact via friction. The interface between the mass and the base can be planar, curved, or have some other complex geometry. The goal of a slope stability analysis is to determine the conditions under which the mass will slip relative to the base and lead to slope failure.

If the interface between the mass and the base of a slope has a complex geometry, slope stability analysis is difficult and numerical solution methods are required. Typically, the exact geometry of the interface is not known and a simplified interface geometry is assumed. Finite slopes require three-dimensional models to be analyzed. To keep the problem simple, most slopes are analyzed assuming that the slopes are infinitely wide and can therefore be represented by two-dimensional models. A slope can be drained or undrained. The undrained condition is used in the calculations to produce conservative estimates of risk.

A popular stability analysis approach is based on principles pertaining to the limit equilibrium concept. This method analyzes a finite or infinite slope as if it were about to fail along its sliding failure surface. Equilibrium stresses are calculated along the failure plane, and compared to the soils shear strength as determined by Terzaghi's shear strength equation. Stability is ultimately decided by a factor of safety equal to the ratio of shear strength to the equilibrium stresses along the failure surface. A factor of safety greater than one generally implies a stable slope, failure of which should not occur assuming the slope is undisturbed. A factor of safety of 1.5 for static conditions is commonly used in practice.

Offshore Geotechnical Engineering

Offshore (or *marine*) *geotechnical engineering* is concerned with foundation design for human-made structures in the sea, away from the coastline (in opposition to *onshore* or *nearshore*). Oil platforms, artificial islands and submarine pipelines are examples of such structures. There are number of significant differences between onshore and offshore geotechnical engineering. Notably, ground improvement (on the seabed) and site investigation are more expensive, the offshore structures are exposed to a wider range of geohazards, and the environmental and financial consequences are higher in case of failure. Offshore structures are exposed to various environmental loads, notably wind, waves and currents. These phenomena may affect the integrity or the serviceability of the structure and its foundation during its operational lifespan – they need to be taken into account in offshore design.

Platforms offshore Mexico.

In subsea geotechnical engineering, seabed materials are considered a two-phase material composed of 1) rock or mineral particles and 2) water. Structures may be fixed in place in the seabed—as is the case for piers, jettys and fixed-bottom wind turbines— or may be a floating structure that remain roughly fixed relative to its geotechnical anchor point. Undersea mooring of human-engineered floating structures include a large number of offshore oil and gas platforms and, since 2008, a few floating wind turbines. Two common types of engineered design for anchoring floating structures include tension-leg and catenary loose mooring systems. "Tension leg mooring systems have vertical tethers under tension providing large restoring moments in pitch and roll. Catenary mooring systems provide station keeping for an offshore structure yet provide little stiffness at low tensions."

Geosynthetics

A collage of geosynthetic products.

Geosynthetics are a type of plastic polymer products used in geotechnical engineering that improve engineering performance while reducing costs. This includes geotextiles, geogrids, geomembranes, geocells, and geocomposites. The synthetic nature of the products make them suitable for use in the ground where high levels of durability are required; their main functions include: drainage, filtration, reinforcement, separation and containment. Geosynthetics are available in a wide range of forms and materials, each to suit a slightly different end use, although they are frequently used together. These products have a wide range of applications and are currently used in many civil and geotechnical engineering applications including: roads, airfields, railroads, embankments, piled embankments, retaining structures, reservoirs, canals, dams, landfills, bank protection and coastal engineering.

Civil Engineering

A multi-level stack interchange, buildings, houses, and park in Shanghai, China.

Civil engineering is a professional engineering discipline that deals with the design, construction, and maintenance of the physical and naturally built environment, including works like roads, bridges, canals, dams, and buildings. Civil engineering is the second-oldest engineering discipline after military engineering, and it is defined to distinguish non-military engineering from military engineering. It is traditionally broken into several sub-disciplines including architectural engineering, environmental engineering, geotechnical engineering, control engineering, structural engineering, earthquake engineering, transportation engineering, forensic engineering, municipal or urban engineering, water resources engineering, materials engineering, wastewater engineering, offshore engineering, facade engineering, coastal engineering, construction surveying, and construction engineering. Civil engineering takes place in the public sector from municipal through to national governments, and in the private sector from individual homeowners through to international companies.

History of the Civil Engineering Profession

Engineering has been an aspect of life since the beginnings of human existence. The earliest practice of civil engineering may have commenced between 4000 and 2000 BC in Ancient Egypt, the Indus Valley Civilization, and Mesopotamia (Ancient Iraq) when humans started to abandon a nomadic existence, creating a need for the construction of shelter. During this time, transportation became increasingly important leading to the development of the wheel and sailing.

Leonhard Euler developed the theory explaining the buckling of columns

Until modern times there was no clear distinction between civil engineering and architecture, and the term engineer and architect were mainly geographical variations referring to the same occupation, and often used interchangeably. The construction of pyramids in Egypt (circa 2700–2500 BC) were some of the first instances of large structure constructions. Other ancient historic civil engineering constructions include the Qanat water management system (the oldest is older than 3000 years and longer than 71 km,) the Parthenon by Iktinos in Ancient Greece (447–438 BC), the Appian Way by Roman engineers (c. 312 BC), the Great Wall of China by General Meng T'ien under orders from Ch'in Emperor Shih Huang Ti (c. 220 BC) and the stupas constructed in ancient Sri Lanka like the Jetavanaramaya and the extensive irrigation works in Anuradhapura. The Romans developed civil structures throughout their empire, including especially aqueducts, insulae, harbors, bridges, dams and roads.

John Smeaton, the "father of civil engineering"

In the 18th century, the term civil engineering was coined to incorporate all things civilian as opposed to military engineering. The first self-proclaimed civil engineer was John Smeaton, who constructed the Eddystone Lighthouse. In 1771 Smeaton and some of his colleagues formed the Smeatonian Society of Civil Engineers, a group of leaders of the profession who met informally over dinner. Though there was evidence of some technical meetings, it was little more than a social society.

In 1818 the Institution of Civil Engineers was founded in London, and in 1820 the eminent engineer Thomas Telford became its first president. The institution received a Royal Charter in 1828, formally recognising civil engineering as a profession. Its charter defined civil engineering as:

the art of directing the great sources of power in nature for the use and convenience of man, as the means of production and of traffic in states, both for external and internal trade, as applied in the construction of roads, bridges, aqueducts, canals, river navigation and docks for internal intercourse and exchange, and in the construction of ports, harbours, moles, breakwaters and lighthouses, and in the art of navigation by artificial power for the purposes of commerce, and in the construction and application of machinery, and in the drainage of cities and towns.

History of Civil Engineering Education

The first private college to teach civil engineering in the United States was Norwich University, founded in 1819 by Captain Alden Partridge. The first degree in civil engineering in the United States was awarded by Rensselaer Polytechnic Institute in 1835. The first such degree to be awarded to a woman was granted by Cornell University to Nora Stanton Blatch in 1905.

In the UK during the early 19th century, the division between civil engineering and military engineering (served by the Royal Military Academy, Woolwich), coupled with the demands of the Industrial Revolution, spawned new engineering education initiatives: the Royal Polytechnic Institution was founded in 1838, the private College for Civil Engineers in Putney was established in 1839, and the UK's first Chair of Engineering was established at the University of Glasgow in 1840.

History of Civil Engineering

Civil engineering is the application of physical and scientific principles for solving the problems of society, and its history is intricately linked to advances in understanding of physics and mathematics throughout history. Because civil engineering is a wide ranging profession, including several separate specialized sub-disciplines, its history is linked to knowledge of structures, materials science, geography, geology, soils, hydrology, environment, mechanics and other fields.

Throughout ancient and medieval history most architectural design and construction was carried out by artisans, such as stonemasons and carpenters, rising to the role of master builder. Knowledge was retained in guilds and seldom supplanted by advances. Structures, roads and infrastructure that existed were repetitive, and increases in scale were incremental.

Chichen Itza was a large pre-Columbian city in Mexico built by the Maya people of the Post Classic. The northeast column temple also covers a channel that funnels all the rainwater from the complex some 40 metres (130 ft) away to a rejollada, a former cenote.

A Roman aqueduct [built circa 19 BC] near Pont du Gard, France

One of the earliest examples of a scientific approach to physical and mathematical problems applicable to civil engineering is the work of Archimedes in the 3rd century BC, including Archimedes Principle, which underpins our understanding of buoyancy, and practical solutions such as Archimedes' screw. Brahmagupta, an Indian mathematician, used arithmetic in the 7th century AD, based on Hindu-Arabic numerals, for excavation (volume) computations.

The Civil Engineer

Education and Licensure

Civil engineers typically possess an academic degree in civil engineering. The length of study is three to five years, and the completed degree is designated as a bachelor of engineering, or a bachelor of science. The curriculum generally includes classes in physics, mathematics, project management, design and specific topics in civil engineering. After taking basic courses in most sub-disciplines of civil engineering, they move onto specialize in one or more sub-disciplines at advanced levels. While an undergraduate degree (BEng/BSc) normally provides successful students with industry-accredited qualification, some academic institutions offer post-graduate degrees (MEng/MSc), which allow students to further specialize in their particular area of interest.

Surveying students with professor at the Helsinki University
of Technology in the late 19th century.

In most countries, a bachelor's degree in engineering represents the first step towards professional certification, and a professional body certifies the degree program. After completing a certified degree program, the engineer must satisfy a range of requirements (including work experience and exam requirements) before being certified. Once certified, the engineer is designated as a professional engineer (in the United States, Canada and South Africa), a chartered engineer (in most Commonwealth countries), a chartered professional engineer (in Australia and New Zealand), or a European engineer (in most countries of the European Union). There are international agreements between relevant professional bodies to allow engineers to practice across national borders.

The benefits of certification vary depending upon location. For example, in the United States and Canada, "only a licensed professional engineer may prepare, sign and seal, and submit engineering plans and drawings to a public authority for approval, or seal engineering work for public and private clients." This requirement is enforced under provincial law such as the Engineers Act in Quebec.

No such legislation has been enacted in other countries including the United Kingdom. In Australia, state licensing of engineers is limited to the state of Queensland. Almost all certifying bodies maintain a code of ethics which all members must abide by.

Engineers must obey contract law in their contractual relationships with other parties. In cases where an engineer's work fails, he may be subject to the law of tort of negligence, and in extreme cases, criminal charges. An engineer's work must also comply with numerous other rules and regulations such as building codes and environmental law.

Sub-disciplines

In general, civil engineering is concerned with the overall interface of human created fixed projects with the greater world. General civil engineers work closely with surveyors and specialized civil engineers to design grading, drainage, pavement, water supply, sewer service,dams, electric and communications supply. General civil engineering is also referred to as site engineering, a branch of civil engineering that primarily focuses on converting a tract of land from one usage to another. Site engineers spend time visiting project sites, meeting with stakeholders, and preparing construction plans. Civil engineers apply the principles of geotechnical engineering, structural engineering, environmental engineering, transportation engineering and construction engineering to residential, commercial, industrial and public works projects of all sizes and levels of construction.

The Akashi Kaikyō Bridge in Japan, currently the world's longest suspension span.

Materials Science and Engineering

Materials science is closely related to civil engineering. It studies fundamental characteristics of materials, and deals with ceramics such as concrete and mix asphalt concrete, strong metals such as aluminum and steel, and polymers including polymethylmethacrylate (PMMA) and carbon fibers.

Materials engineering involves protection and prevention (paints and finishes). Alloying combines two types of metals to produce another metal with desired properties. It incorporates elements of applied physics and chemistry. With recent media attention on nanoscience and nanotechnology, materials engineering has been at the forefront of academic research. It is also an important part of forensic engineering and failure analysis.

Coastal Engineering

Coastal engineering is concerned with managing coastal areas. In some jurisdictions, the terms sea defense and coastal protection mean defense against flooding and erosion, respectively. The term coastal defense is the more traditional term, but coastal management has become more popular as the field has expanded to techniques that allow erosion to claim land.

Oosterscheldekering, a storm surge barrier in the Netherlands.

Construction Engineering

Construction engineering involves planning and execution, transportation of materials, site development based on hydraulic, environmental, structural and geotechnical engineering. As construction firms tend to have higher business risk than other types of civil engineering firms do, construction engineers often engage in more business-like transactions, for example, drafting and reviewing contracts, evaluating logistical operations, and monitoring prices of supplies.

Earthquake Engineering

Earthquake engineering involves designing structures to withstand hazardous earthquake exposures. Earthquake engineering is a sub-discipline of structural engineering. The main objectives of earthquake engineering are to understand interaction of structures on the shaky ground; foresee the consequences of possible earthquakes; and design, construct and maintain structures to perform at earthquake in compliance with building codes.

Environmental Engineering

Water pollution

Environmental engineering is the contemporary term for sanitary engineering, though sanitary engineering traditionally had not included much of the hazardous waste management and environmental remediation work covered by environmental engineering. Public health engineering and environmental health engineering are other terms being used.

Environmental engineering deals with treatment of chemical, biological, or thermal wastes, purification of water and air, and remediation of contaminated sites after waste disposal or accidental contamination. Among the topics covered by environmental engineering are pollutant transport, water purification, waste water treatment, air pollution, solid waste treatment, and hazardous waste management. Environmental engineers administer pollution reduction, green engineering, and industrial ecology. Environmental engineers also compile information on environmental consequences of proposed actions.

Geotechnical Engineering

Geotechnical engineering studies rock and soil supporting civil engineering systems. Knowledge from the field of soil science, materials science, mechanics, and hydraulics is applied to safely and economically design foundations, retaining walls, and other structures. Environmental efforts to protect groundwater and safely maintain landfills have spawned a new area of research called geoenvironmental engineering.

Identification of soil properties presents challenges to geotechnical engineers. Boundary conditions are often well defined in other branches of civil engineering, but unlike steel or concrete, the material properties and behavior of soil are difficult to predict due to its variability and limitation on investigation. Furthermore, soil exhibits nonlinear (stress-dependent) strength, stiffness, and dilatancy (volume change associated with application of shear stress), making studying soil mechanics all the more difficult.

Water Resources Engineering

Water resources engineering is concerned with the collection and management of water (as a natural resource). As a discipline it therefore combines elements of hydrology, environmental science, meteorology, conservation, and resource management. This area of civil engineering relates to the prediction and management of both the quality and the quantity of water in both underground (aquifers) and above ground (lakes, rivers, and streams) resources. Water resource engineers analyze and model very small to very large areas of the earth to predict the amount and content of water as it flows into, through, or out of a facility. Although the actual design of the facility may be left to other engineers.

Hydraulic engineering is concerned with the flow and conveyance of fluids, principally water. This area of civil engineering is intimately related to the design of pipelines, water supply network, drainage facilities (including bridges, dams, channels, culverts, levees, storm sewers), and canals. Hydraulic engineers design these facilities using the concepts of fluid pressure, fluid statics, fluid dynamics, and hydraulics, among others.

Structural Engineering

Shallow foundation construction example

Structural engineering is concerned with the structural design and structural analysis of buildings, bridges, towers, flyovers (overpasses), tunnels, off shore structures like oil and gas fields in the sea, aerostructure and other structures. This involves identifying the loads which act upon a structure and the forces and stresses which arise within that structure due to those loads, and then designing the structure to successfully support and resist those loads. The loads can be self weight of the structures, other dead load, live loads, moving (wheel) load, wind load, earthquake load, load from temperature change etc. The structural engineer must design structures to be safe for their users and to successfully fulfill the function they are designed for (to be *serviceable*). Due to the nature of some loading conditions, sub-disciplines within structural engineering have emerged, including wind engineering and earthquake engineering.

Design considerations will include strength, stiffness, and stability of the structure when subjected to loads which may be static, such as furniture or self-weight, or dynamic, such as wind, seismic, crowd or vehicle loads, or transitory, such as temporary construction loads or impact. Other considerations include cost, constructability, safety, aesthetics and sustainability.

Surveying

Surveying is the process by which a surveyor measures certain dimensions that occur on or near the surface of the Earth. Surveying equipment, such as levels and theodolites, are used for accurate measurement of angular deviation, horizontal, vertical and slope distances. With computerisation, electronic distance measurement (EDM), total stations, GPS surveying and laser scanning have to a large extent supplanted traditional instruments. Data collected by survey measurement is converted into a graphical representation of the Earth's surface in the form of a map. This information is then used by civil engineers, contractors and realtors to design from, build on, and trade, respectively. Elements of a structure must be sized and positioned in relation to each other and to site boundaries and adjacent structures. Although surveying is a distinct profession with separate qualifications and licensing arrangements, civil engineers are trained in the basics of surveying and mapping, as well as geographic information systems. Surveyors also lay out the routes of railways, tramway tracks, highways, roads, pipelines and streets as well as position other infrastructure, such as harbors, before construction.

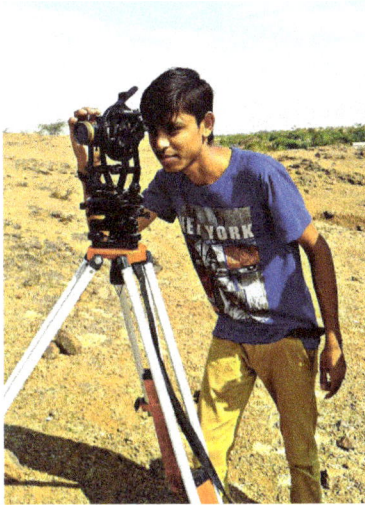

Civil engineering student using a theodolite.

Land Surveying

BLM cadastral survey marker from 1992 in San Xavier, Arizona.

In the United States, Canada, the United Kingdom and most Commonwealth countries land surveying is considered to be a separate and distinct profession. Land surveyors are not considered to be engineers, and have their own professional associations and licensing requirements. The services of a licensed land surveyor are generally required for boundary surveys (to establish the boundaries of a parcel using its legal description) and subdivision plans (a plot or map based on a survey of a parcel of land, with boundary lines drawn inside the larger parcel to indicate the creation of new boundary lines and roads), both of which are generally referred to as Cadastral surveying.

Construction Surveying

Construction surveying is generally performed by specialised technicians. Unlike land surveyors, the resulting plan does not have legal status. Construction surveyors perform the following tasks:

- Surveying existing conditions of the future work site, including topography, existing buildings and infrastructure, and underground infrastructure when possible;

- "lay-out" or "setting-out": placing reference points and markers that will guide the construction of new structures such as roads or buildings;

- Verifying the location of structures during construction;

- As-Built surveying: a survey conducted at the end of the construction project to verify that the work authorized was completed to the specifications set on plans.

Transportation Engineering

The engineering of this roundabout in Bristol, England, attempts to make traffic flow free-moving

Transportation engineering is concerned with moving people and goods efficiently, safely, and in a manner conducive to a vibrant community. This involves specifying, designing, constructing, and maintaining transportation infrastructure which includes streets, canals, highways, rail systems, airports, ports, and mass transit. It includes areas such as transportation design, transportation planning, traffic engineering, some aspects of urban engineering, queueing theory, pavement engineering, Intelligent Transportation System (ITS), and infrastructure management.

Forensic Engineering

Forensic engineering is the investigation of materials, products, structures or components that fail or do not operate or function as intended, causing personal injury or damage to property. The consequences of failure are dealt with by the law of product liability. The field also deals with retracing processes and procedures leading to accidents in operation of vehicles or machinery. The subject is applied most commonly in civil law cases, although it may be of use in criminal law cases. Generally the purpose of a Forensic engineering investigation is to locate cause or causes of failure with a view to improve performance or life of a component, or to assist a court in determining the facts of an accident. It can also involve investigation of intellectual property claims, especially patents.

Municipal or Urban Engineering

Lake Chapultepec

Municipal engineering is concerned with municipal infrastructure. This involves specifying, designing, constructing, and maintaining streets, sidewalks, water supply networks, sewers, street lighting, municipal solid waste management and disposal, storage depots for various bulk materials used for maintenance and public works (salt, sand, etc.), public parks and cycling infrastructure. In the case of underground utility networks, it may also include the civil portion (conduits and access chambers) of the local distribution networks of electrical and telecommunications services. It can also include the optimizing of waste collection and bus service networks. Some of these disciplines overlap with other civil engineering specialties, however municipal engineering focuses on the coordination of these infrastructure networks and services, as they are often built simultaneously, and managed by the same municipal authority. Municipal engineers may also design the site civil works for large buildings, industrial plants or campuses (i.e. access roads, parking lots, potable water supply, treatment or pretreatment of waste water, site drainage, etc.)

Control Engineering

Control engineering (or *control systems engineering*) is the branch of civil engineering discipline that applies control theory to design systems with desired behaviors. The practice uses sensors to measure the output performance of the device being controlled (often a vehicle) and those measurements can be used to give feedback to the input actuators that can make corrections toward desired performance. When a device is designed to perform without the need of human inputs for correction it is called automatic control (such as cruise control for regulating a car's speed). Multidisciplinary in nature, control systems engineering activities focus on implementation of control systems mainly derived by mathematical modeling of systems of a diverse range.

Civil Engineering Associations

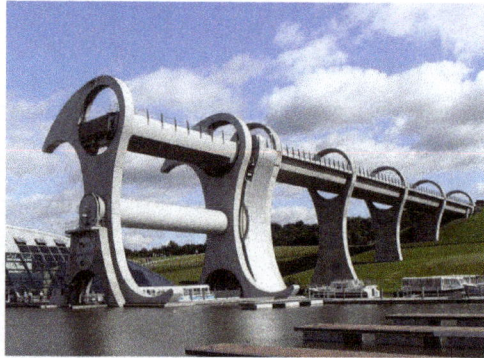

The Falkirk Wheel in Scotland

- American Society of Civil Engineers

- Canadian Society for Civil Engineering

- Conselho Federal de Engenharia e Agronomia (pt)

- Earthquake Engineering Research Institute

- Engineers Australia

- European Federation of National Engineering Associations

- International Federation of Consulting Engineers

- Institution of Civil Engineers

- The Institution of Civil Engineering Surveyors

- Institution of Engineers (India)

- Institution of Engineers of Ireland

- Institute of Transportation Engineers

- Pakistan Engineering Council

- Philippine Institute of Civil Engineers

- Transportation Research Board

Seismic Analysis

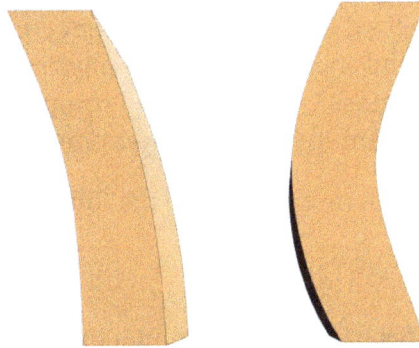

First and second modes of building seismic response

Seismic analysis is a subset of structural analysis and is the calculation of the response of a building (or nonbuilding) structure to earthquakes. It is part of the process of structural design, earthquake engineering or structural assessment and retrofit in regions where earthquakes are prevalent.

As seen in the figure, a building has the potential to 'wave' back and forth during an earthquake (or even a severe wind storm). This is called the 'fundamental mode', and is the lowest frequency of building response. Most buildings, however, have higher modes of response, which are uniquely activated during earthquakes. The figure just shows the second mode, but there are higher 'shimmy' (abnormal vibration) modes. Nevertheless, the first and second modes tend to cause the most damage in most cases.

The earliest provisions for seismic resistance were the requirement to design for a lateral force equal to a proportion of the building weight (applied at each floor level). This approach was adopted in the appendix of the 1927 Uniform Building Code (UBC), which was used on the west coast of the United States. It later became clear that the dynamic properties of the structure affected the loads generated during an earthquake. In the Los Angeles County Building Code of 1943 a provision to vary the load based on the number of floor levels was adopted (based on research carried out at Caltech in collaboration with Stanford University and the U.S. Coast and Geodetic Survey, which started in 1937). The concept of "response spectra" was developed in the 1930s, but it wasn't until 1952 that a joint committee of the San Francisco Section of the ASCE and the Structural Engineers Association of Northern California (SEAONC) proposed using the building period (the inverse of the frequency) to determine lateral forces.

The University of California, Berkeley was an early base for computer-based seismic analysis of structures, led by Professor Ray Clough (who coined the term finite element). Students included Ed Wilson, who went on to write the program SAP in 1970, an early "Finite Element Analysis" program.

Earthquake engineering has developed a lot since the early days, and some of the more complex designs now use special earthquake protective elements either just in the foundation (base isolation) or distributed throughout the structure. Analyzing these types of structures requires specialized explicit finite element computer code, which divides time into very small slices and models the actual physics, much like common video games often have "physics engines". Very large and complex buildings can be modeled in this way (such as the Osaka International Convention Center).

Structural analysis methods can be divided into the following five categories.

Equivalent Static Analysis

This approach defines a series of forces acting on a building to represent the effect of earthquake ground motion, typically defined by a seismic design response spectrum. It assumes that the building responds in its fundamental mode. For this to be true, the building must be low-rise and must not twist significantly when the ground moves. The response is read from a design response spectrum, given the natural frequency of the building (either calculated or defined by the building code). The applicability of this method is extended in many building codes by applying factors to account for higher buildings with some higher modes, and for low levels of twisting. To account for effects due to "yielding" of the structure, many codes apply modification factors that reduce the design forces (e.g. force reduction factors).

Response Spectrum Analysis

This approach permits the multiple modes of response of a building to be taken into account (in the frequency domain). This is required in many building codes for all except for very simple or very complex structures. The response of a structure can be defined as a combination of many special shapes (modes) that in a vibrating string correspond to the "harmonics". Computer analysis can be used to determine these modes for a structure. For each mode, a response is read from the design spectrum, based on the modal frequency and the modal mass, and they are then combined to provide an estimate of the total response of the structure. In this we have to calculate the magnitude of forces in all directions i.e. X, Y & Z. Combination methods include the following:

- absolute - peak values are added together

- square root of the sum of the squares (SRSS)

- complete quadratic combination (CQC) - a method that is an improvement on SRSS for closely spaced modes

The result of a response spectrum analysis using the response spectrum from a ground motion is typically different from that which would be calculated directly from a linear dynamic analysis using that ground motion directly, since phase information is lost in the process of generating the response spectrum.

In cases where structures are either too irregular, too tall or of significance to a community in disaster response, the response spectrum approach is no longer appropriate, and more complex analysis is often required, such as non-linear static analysis or dynamic analysis.

Linear Dynamic Analysis

Static procedures are appropriate when higher mode effects are not significant. This is generally true for short, regular buildings. Therefore, for tall buildings, buildings with torsional irregularities, or non-orthogonal systems, a dynamic procedure is required. In the linear dynamic procedure, the building is modelled as a multi-degree-of-freedom (MDOF) system with a linear elastic stiffness matrix and an equivalent viscous damping matrix.

The seismic input is modelled using either modal spectral analysis or time history analysis but in both cases, the corresponding internal forces and displacements are determined using linear elastic analysis. The advantage of these linear dynamic procedures with respect to linear static procedures is that higher modes can be considered. However, they are based on linear elastic response and hence the applicability decreases with increasing nonlinear behaviour, which is approximated by global force reduction factors.

In linear dynamic analysis, the response of the structure to ground motion is calculated in the time domain, and all phase information is therefore maintained. Only linear properties are assumed. The analytical method can use modal decomposition as a means of reducing the degrees of freedom in the analysis.

Nonlinear Static Analysis

In general, linear procedures are applicable when the structure is expected to remain nearly elastic for the level of ground motion or when the design results in nearly uniform distribution of nonlinear response throughout the structure. As the performance objective of the structure implies greater inelastic demands, the uncertainty with linear procedures increases to a point that requires a high level of conservatism in demand assumptions and acceptability criteria to avoid unintended performance. Therefore, procedures incorporating inelastic analysis can reduce the uncertainty and conservatism.

This approach is also known as "pushover" analysis. A pattern of forces is applied to a structural model that includes non-linear properties (such as steel yield), and the total force is plotted against a reference displacement to define a capacity curve. This can then be combined with a demand curve (typically in the form of an acceleration-displacement response spectrum (ADRS)). This essentially reduces the problem to a single degree of freedom (SDOF) system.

Nonlinear static procedures use equivalent SDOF structural models and represent seismic ground motion with response spectra. Story drifts and component actions are related subsequently to the global demand parameter by the pushover or capacity curves that are the basis of the non-linear static procedures.

Nonlinear Dynamic Analysis

Nonlinear dynamic analysis utilizes the combination of ground motion records with a detailed structural model, therefore is capable of producing results with relatively low uncertainty. In nonlinear dynamic analyses, the detailed structural model subjected to a ground-motion record produces estimates of component deformations for each degree of freedom in the model and the modal responses are combined using schemes such as the square-root-sum-of-squares.

In non-linear dynamic analysis, the non-linear properties of the structure are considered as part of a time domain analysis. This approach is the most rigorous, and is required by some building codes for buildings of unusual configuration or of special importance. However, the calculated response can be very sensitive to the characteristics of the individual ground motion used as seismic input; therefore, several analyses are required using different ground motion records to achieve a reliable estimation of the probabilistic distribution of structural response. Since the properties of the seismic response depend on the intensity, or severity, of the seismic shaking, a comprehensive assessment calls for numerous nonlinear dynamic analyses at various levels of intensity to represent different possible earthquake scenarios. This has led to the emergence of methods like the Incremental Dynamic Analysis.

References

- Blank, Alan; McEvoy, Michael; Plank, Roger (1993). Architecture and Construction in Steel. Taylor & Francis. ISBN 0-419-17660-8.

- Hewson, Nigel R. (2003). Prestressed Concrete Bridges: Design and Construction. Thomas Telford. ISBN 0-7277-2774-5.

- Terzaghi, K., Peck, R.B. and Mesri, G. (1996), Soil Mechanics in Engineering Practice 3rd Ed., John Wiley & Sons, Inc. ISBN 0-471-08658-4.

- Holtz, R. and Kovacs, W. (1981), An Introduction to Geotechnical Engineering, Prentice-Hall, Inc. ISBN 0-13-484394-0.

- Budhu, Muni (2007). Soil Mechanics and Foundations. John Wiley & Sons, Inc. ISBN 978-0-471-43117-6.

- Soil Mechanics, Lambe,T.William and Whitman,Robert V., Massachusetts Institute of Technology, John Wiley & Sons., 1969. ISBN 0-471-51192-7

- Soil Behavior and Critical State Soil Mechanics, Wood, David Muir,Cambridge University Press, 1990. ISBN 0-521-33782-8

- Coduto, Donald; et al. (2011). Geotechnical Engineering Principles and Practices. New Jersey: Pearson Higher Education. ISBN 9780132368681.

- RAJU, V. R. (2010). Ground Improvement Technologies and Case Histories. Singapore: Research Publishing Services. p. 809. ISBN 978-981-08-3124-0.

- "ETABS receives "Top Seismic Product of the 20th Century" Award" (PDF). Press Release. Structure Magazine. 2006. Retrieved April 20, 2012.

Essential Aspects of Earthquake Engineering

Disaster management or emergency management closely considers contingency planning and decreasing the damage caused by disasters. It seeks to provide information as well as material for effective responses to disasters. This chapter studies peak ground acceleration and emergency management in an effort to comprehensively study the essential aspects of earthquake engineering.

Emergency Management

Disaster management (or emergency management) is the creation of plans through which communities reduce vulnerability to hazards and cope with disasters. Disaster management does not avert or eliminate the threats; instead, it focuses on creating plans to decrease the effect of disasters. Failure to create a plan could lead to human mortality, lost revenue, and damage to assets. Currently in the United States 60 percent of businesses do not have emergency management plans. Events covered by disaster management include acts of terrorism, industrial sabotage, fire,natural disasters (such as earthquakes, hurricanes, etc.), public disorder, industrial accidents, and communication failures.

Emergency Planning Ideals

If possible, emergency planning should aim to prevent emergencies from occurring, and failing that, should develop a good action plan to mitigate the results and effects of any emergencies. As time goes on, and more data becomes available, usually through the study of emergencies as they occur, a plan should evolve. The development of emergency plans is a cyclical process, common to many risk management disciplines, such as Business Continuity and Security Risk Management, as set out below:

- Recognition or identification of risks
- Ranking or evaluation of risks
    - Responding to significant risks
    - Tolerate
    - Treat

- o Transfer

- o Terminate

- Resourcing controls

- Reaction Planning

- Reporting & monitoring risk performance

- Reviewing the Risk Management framework

There are a number of guidelines and publications regarding Emergency Planning, published by various professional organizations such as ASIS, National Fire Protection Association (NFPA), and the International Association of Emergency Managers (IAEM). There are very few Emergency Management specific standards, and emergency management as a discipline tends to fall under business resilience standards.

In order to avoid, or reduce significant losses to a business, emergency managers should work to identify and anticipate potential risks, hopefully to reduce their probability of occurring. In the event that an emergency does occur, managers should have a plan prepared to mitigate the effects of that emergency, as well as to ensure Business Continuity of critical operations post-incident. It is essential for an organisation to include procedures for determining whether an emergency situation has occurred and at what point an emergency management plan should be activated.

Implementation Ideals

An emergency plan must be regularly maintained, in a structured and methodical manner, to ensure it is up-to-date in the event of an emergency. Emergency managers generally follow a common process to anticipate, assess, prevent, prepare, respond and recover from an incident.

Pre-incident Training and Testing

A team of emergency responders performs a training scenario involving anthrax.

Emergency management plans and procedures should include the identification of appropriately trained staff members responsible for decision-making when an emergency occurs. Training plans should include internal people, contractors and civil protection partners, and should state the nature and frequency of training and testing.

Testing of a plan's effectiveness should occur regularly. In instances where several business or organisations occupy the same space, joint emergency plans, formally agreed to by all parties, should be put into place.

Communicating and Incident Assessment

Communication is one of the key issues during any emergency, pre-planning of communications is critical. Miscommunication can easily result in emergency events escalating unnecessarily.

Once an emergency has been identified a comprehensive assessment evaluating the level of impact and its financial implications should be undertaken. Following assessment, the appropriate plan or response to be activated will depend on a specific pre-set criteria within the emergency plan. The steps necessary should be prioritized to ensure critical functions are operational as soon as possible.

Phases and Personal Activities

Emergency management consists of five phases: prevention, mitigation, preparedness, response and recovery.

Prevention

Prevention was recently added to the phases of emergency management. It focuses on preventing the human hazard, primarily from potential natural disasters or terrorist attacks. Preventive measures are taken on both the domestic and international levels, designed to provide permanent protection from disasters. Not all disasters, particularly natural disasters, can be prevented, but the risk of loss of life and injury can be mitigated with good evacuation plans, environmental planning and design standards. In January 2005, 167 Governments adopted a 10-year global plan for natural disaster risk reduction called the Hyogo Framework.

Preventing or reducing the impacts of disasters on our communities is a key focus for emergency management efforts today. Prevention and mitigation also help reduce the financial costs of disaster response and recovery. Public Safety Canada is working with provincial and territorial governments and stakeholders to promote disaster prevention and mitigation using a risk-based and all-hazards approach. In 2008, Federal/Provincial/Territorial Ministers endorsed a National Disaster Mitigation Strategy.

Mitigation

Preventive or mitigation measures take different forms for different types of disasters. In earthquake prone areas, these preventive measures might include structural changes such as the installation of an earthquake valve to instantly shut off the natural gas supply, seismic retrofits of property, and the securing of items inside a building. The latter may include the mounting of furniture, refrigerators, water heaters and breakables to the walls, and the addition of cabinet latches. In flood prone areas, houses can be built on poles/stilts. In areas prone to prolonged electricity black-outs installation of a generator ensures continuation of electrical service. The construction of storm cellars and fallout shelters are further examples of personal mitigative actions.

On a national level, governments might implement large scale mitigation measures. After the monsoon floods of 2010, the Punjab government subsequently constructed 22 'disaster-resilient' model villages, comprising 1885 single-storey homes, together with schools and health centres.

Disaster mitigation measures are those that eliminate or reduce the impacts and risks of hazards through proactive measures taken before an emergency or disaster occurs.

One of the best known examples of investment in disaster mitigation is the Red River Floodway. The building of the Floodway was a joint provincial/federal undertaking to protect the City of Winnipeg and reduce the impact of flooding in the Red River Basin. It cost $60 million to build in the 1960s. Since then, the floodway has been used over 20 times. Its use during the 1997 Red River Flood alone saved an estimated $6 billion. The Floodway was expanded in 2006 as a joint provincial/federal initiative.

Preparedness

An airport emergency preparedness exercise

Preparedness focuses on preparing equipment and procedures for use when a disaster occurs. This equipment and these procedures can be used to reduce vulnerability to disaster, to mitigate the impacts of a disaster or to respond more efficiently in an emergency. The Federal Emergency Management Agency (FEMA) has set out a basic four-stage vision of preparedness flowing from mitigation to preparedness to response to recovery and back to mitigation in a circular planning process. This circular, overlapping model has been modified by other agencies, taught in emergency class and discussed in academic papers. FEMA also operates a Building Science Branch that develops and produces multi-hazard mitigation guidance that focuses on creating disaster-resilient communities to reduce loss of life and property. FEMA advises citizens to prepare their homes with some emergency essentials in the case that the food distribution lines are interrupted. FEMA has subsequently prepared for this contingency by purchasing hundreds of thousands of freeze dried food emergency meals ready to eat (MRE's) to dispense to the communities where emergency shelter and evacuations are implemented.

Emergency preparedness can be difficult to measure. CDC focuses on evaluating the effectiveness of its public health efforts through a variety of measurement and assessment programs.

Local Emergency Planning Committees

Local Emergency Planning Committees (LEPCs) are required by the United States Environmental Protection Agency under the Emergency Planning and Community Right-to-Know Act to develop an emergency response plan, review the plan at least annually, and provide information about chemicals in the community to local citizens. This emergency preparedness effort focuses on hazards presented by use and storage of extremely hazardous, hazardous and toxic chemicals. Particular requirements of LEPCs include

- Identification of facilities and transportation routes of extremely hazardous substances

- Description of emergency response procedures, on and off site

- Designation of a community coordinator and facility emergency coordinator(s) to implement the plan

- Outline of emergency notification procedures

- Description of how to determine the probable affected area and population by releases

- Description of local emergency equipment and facilities and the persons responsible for them

- Outline of evacuation plans

- A training program for emergency responders (including schedules)

- Methods and schedules for exercising emergency response plans

According to the EPA, "Many LEPCs have expanded their activities beyond the requirements of EPCRA, encouraging accident prevention and risk reduction, and addressing homeland security in their communities" and the Agency offers advice on how to evaluate the effectiveness of these committees.

Preparedness Measures

Preparedness measures can take many forms ranging from focusing on individual people, locations or incidents to broader, government-based "all hazard" planning. There are a number of preparedness stages between "all hazard' and individual planning, generally involving some combination of both mitigation and response planning. Business continuity planning encourages businesses to have a Disaster Recovery Plan. Community- and faith-based organizations mitigation efforts promote field response teams and inter-agency planning.

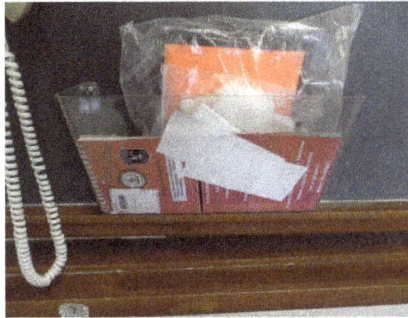

Classroom response kit

School-based response teams cover everything from live shooters to gas leaks and nearby bank robberies. Educational institutions plan for cyberattacks and windstorms. Industry specific guidance exists for horse farms, boat owners and more.

Family preparedness for disaster is fairly unusual. A 2013 survey found that only 19% of American families felt that they were "very prepared" for a disaster. Still, there are many resources available for family disaster planning. The Department of Homeland Security's Ready.gov page includes a Family Emergency Plan Checklist, has a whole webpage devoted to readiness for kids, complete with cartoon-style superheroes, and ran a Thunderclap Campaign in 2014. The Center for Disease Control has a Zombie Apocalypse website.

Kitchen fire extinguisher

Disasters take a variety of forms to include earthquakes, tsunamis or regular structure fires. That a disaster or emergency is not large scale in terms of population or acreage impacted or duration does not make it any less of a disaster for the people or area impacted and much can be learned about preparedness from so-called small disasters. The Red Cross states that it responds to nearly 70,000 disasters a year, the most common of which is a single family fire.

Items on shelves in basement

Preparedness starts with an individual's everyday life and involves items and training that would be useful in an emergency. What is useful in an emergency is often also useful in everyday life. From personal preparedness, preparedness continues on a continuum through family preparedness, community preparedness and then business, non-profit and governmental preparedness. Some organizations blend these various levels. For example, the International Red Cross and Red Crescent Movement has a webpage on disaster training as well as offering training on basic preparedness such as Cardiopulmonary resuscitation and First Aid. Other non-profits such as Team Rubicon bring specific groups of people into disaster preparedness and response operations. FEMA breaks down preparedness into a pyramid, with citizens on the foundational bottom, on top of which rests local government, state government and federal government in that order.

Non-perishable food in cabinet

The basic theme behind preparedness is to be ready for an emergency and there are a number of different variations of being ready based on an assessment of what sort of threats exist. Nonetheless, there is basic guidance for preparedness that is common despite an area's specific dangers. FEMA recommends that everyone have a three-day survival kit for their household. Because individual household sizes and specific needs might vary, FEMA's recommendations are not item specific, but the list includes:

- Three-day supply of non-perishable food.
- Three-day supply of water – one gallon of water per person, per day.
- Portable, battery-powered radio or television and extra batteries.
- Flashlight and extra batteries.
- First aid kit and manual.
- Sanitation and hygiene items (e.g. toilet paper, menstrual hygiene products).
- Matches and waterproof container.
- Whistle.
- Extra clothing.
- Kitchen accessories and cooking utensils, including a can opener.
- Photocopies of credit and identification cards.
- Cash and coins.
- Special needs items, such as prescription medications, eyeglasses, contact lens solutions, and hearing aid batteries.
- Items for infants, such as formula, diapers, bottles, and pacifiers.
- Other items to meet unique family needs.

Along similar lines, but not exactly the same, CDC has its own list for a proper disaster supply kit.

- Water—one gallon per person, per day
- Food—nonperishable, easy-to-prepare items
- Flashlight
- Battery powered or hand crank radio (NOAA Weather Radio, if possible)
- Extra batteries

- First aid kit

- Medications (7-day supply), other medical supplies, and medical paperwork (e.g., medication list and pertinent medical information)

- Multipurpose tool (e.g., Swiss army knife)

- Sanitation and personal hygiene items

- Copies of personal documents (e.g., proof of address, deed/lease to home, passports, birth certificates, and insurance policies)

- Cell phone with chargers

- Family and emergency contact information

- Extra cash

- Emergency blanket

- Map(s) of the area

- Extra set of car keys and house keys

- Manual can opener

Children are a special population when considering Emergency preparedness and many resources are directly focused on supporting them. SAMHSA has list of tips for talking to children during infectious disease outbreaks, to include being a good listener, encouraging children to ask questions and modeling self-care by setting routines, eating healthy meals, getting enough sleep and taking deep breaths to handle stress. FEMA has similar advice, noting that "Disasters can leave children feeling frightened, confused, and insecure" whether a child has experienced it first hand, had it happen to a friend or simply saw it on television. In the same publication, FEMA further notes, "Preparing for disaster helps everyone in the family accept the fact that disasters do happen, and provides an opportunity to identify and collect the resources needed to meet basic needs after disaster. Preparation helps; when people feel prepared, they cope better and so do children."

To help people assess what threats might be in order to augment their emergency supplies or improve their disaster response skills, FEMA has published a booklet called the "Threat and Hazard Identification and Risk Assessment Guide." (THIRA) This guide, which outlines the THIRA process, emphasizes "whole community involvement," not just governmental agencies, in preparedness efforts. In this guide, FEMA breaks down hazards into three categories: Natural, technological and human caused and notes that each hazard should be assessed for both its likelihood and its significance. According to FEMA, "Communities should consider only those threats and hazards that could plausibly occur" and "Communities should consider only those threats and hazards that would have a significant effect on them." To develop threat and hazard context descriptions, communities should take into account the time, place, and conditions in which threats or hazards might occur.

Not all preparedness efforts and discussions involve the government or established NGOs like the Red Cross. Emergency preparation discussions are active on the internet, with many blogs and websites dedicated to discussing various aspects of preparedness. On-line sales of items such as survival food, medical supplies and heirloom seeds allow people to stock basements with cases of food and drinks with 25 year shelf lives, sophisticated medical kits and seeds that are guaranteed to sprout even after years of storage.

One group of people who put a lot of effort in disaster preparations is called Doomsday Preppers. This subset of preparedness-minded people often share a belief that the FEMA or Red Cross emergency preparation suggestions and training are not extensive enough. Sometimes called survivalists, Doomsday Preppers are often preparing for The End Of The World As We Know It, abbreviated as TEOTWAWKI. With a motto some have that "The Future Belongs to those who Prepare," this Preparedness subset has its own set of Murphy's Rules, including "Rule Number 1: Food, you still don't have enough" and "Rule Number 26: People who thought the Government would save them, found out that it didn't."

Not all emergency preparation efforts revolve around food, guns and shelters, though these items help address the needs in the bottom two sections of Maslow's hierarchy of needs. The American Preppers Network has an extensive list of items that might be useful in less apparent ways than a first aid kid or help add 'fun' to challenging times. These items include:

- Books and magazines
- Arts and crafts painting
- Children's entertainment
- Crayons and coloring books
- Notebooks and writing supplies
- Nuts, bolts, screws, nails, etc.
- Religious material
- Sporting equipment, card games and board games
- Posters and banners creating awareness

Emergency preparedness goes beyond immediate family members. For many people, pets are an integral part of their families and emergency preparation advice includes them as well. It is not unknown for pet owners to die while trying to rescue their pets from a fire or from drowning. CDC's Disaster Supply Checklist for Pets includes:

- Food and water for at least 3 days for each pet; bowls, and a manual can opener.

- Depending on the pet you may need a litter box, paper towels, plastic trash bags, grooming items, and/or household bleach.

- Medications and medical records stored in a waterproof container.

- First aid kit with a pet first aid book.

- Sturdy leash, harness, and carrier to transport pet safely. A carrier should be large enough for the animal to stand comfortably, turn around, and lie down. Your pet may have to stay in the carrier for several hours.

- Pet toys and the pet's bed, if you can easily take it, to reduce stress.

- Current photos and descriptions of your pets to help others identify them in case you and your pets become separated, and to prove that they are yours.

- Information on feeding schedules, medical conditions, behavior problems, and the name and telephone number of your veterinarian in case you have to board your pets or place them in foster care.

Emergency preparedness also includes more than physical items and skill-specific training. Psychological preparedness is also a type of emergency preparedness and specific mental health preparedness resources are offered for mental health professionals by organizations such as the Red Cross. These mental health preparedness resources are designed to support both community members affected by a disaster and the disaster workers serving them. CDC has a website devoted to coping with a disaster or traumatic event. After such an event, the CDC, through the Substance Abuse and Mental Health Services Administration (SAMHSA), suggests that people seek psychological help when they exhibit symptoms such as excessive worry, crying frequently, an increase in irritability, anger, and frequent arguing, wanting to be alone most of the time, feeling anxious or fearful, overwhelmed by sadness, confused, having trouble thinking clearly and concentrating, and difficulty making decisions, increased alcohol and/or substance use, increased physical (aches, pains) complaints such as headaches and trouble with "nerves."

Sometimes emergency supplies are kept in what is called a Bug-out bag. While FEMA does not actually use the term "Bug out bag," calling it instead some variation of a "Go Kit," the idea of having emergency items in a quickly accessible place is common to both FEMA and CDC, though on-line discussions of what items a "bug out bag" should include sometimes cover items such as firearms and great knives that are not specifically suggested by FEMA or CDC. The theory behind a "bug out bag" is that emergency preparations should include the possibility of Emergency evacuation. Whether fleeing a burning building or hastily packing a car to escape an impending hurricane, flood or dangerous chemical release, rapid departure from a home or workplace environment is always a possibility and FEMA suggests having a Family Emergency Plan for such occasions. Because family members may not be together when disaster strikes, this plan should include reliable contact information for friends or relatives who live outside of what would be the disaster area for household members to notify they are safe or otherwise communicate with each other. Along with the contact information, FEMA suggests having well-understood local gathering points if a house must be evacuated quickly to avoid the dangers of re-reentering a burning home. Family and emergency contact information should be printed on cards and put in each family member's backpack or wallet. If family members spend a significant amount of time in a specific location, such as at work or school, FEMA suggests learning the emergency preparation plans for those places. FEMA has a specific form, in English and in Spanish, to help people put together these emergency plans, though it lacks lines for email contact information.

Like children, people with disabilities and other special needs have special emergency preparation needs. While "disability" has a specific meaning for specific organizations such as collecting Social Security benefits, for the purposes of emergency preparedness, the Red Cross uses the term in a broader sense to include people with physical, medical, sensor or cognitive disabilities or the elderly and other special needs populations. Depending on the particular disability, specific emergency preparations might be required. FEMA's suggestions for people with disabilities includes having copies of prescriptions, charging devices for medical devices such as motorized wheel chairs and a week's supply of medication readily available LINK or in a "go stay kit." In some instances, lack of competency in English may lead to special preparation requirements and communication efforts for both individuals and responders.

FEMA notes that long term power outages can cause damage beyond the original disaster that can be mitigated with emergency generators or other power sources to provide an Emergency power system. The United States Department of Energy states that 'homeowners, business owners, and local leaders may have to take an active role in dealing with energy disruptions on their own." This active role may include installing or other procuring generators that are either portable or permanently mounted and run on fuels such as propane or natural gas or gasoline. Concerns about carbon monoxide poisoning, electrocution, flooding, fuel storage and fire lead even small property owners to consider professional installation and maintenance. Major institutions like hospitals, military bases and educational institutions often have or are considering extensive backup power systems. Instead of, or in addition to, fuel-based power systems, solar, wind and other alternative power sources may be used. Standalone batteries, large or small, are also used to provide backup charging for electrical systems and devices ranging from emergency lights to computers to cell phones.

Emergency preparedness does not stop at home or at school. The United States Department of Health and Human Services addresses specific emergency preparedness issues hospitals may have to respond to, including maintaining a safe temperature, providing adequate electricity for life support systems and even carrying out evacuations under extreme circumstances. FEMA encourages all businesses to have businesses to have an emergency response plan and the Small Business Administration specifically advises small business owners to also focus emergency preparedness and provides a variety of different worksheets and resources.

FEMA cautions that emergencies happen while people are travelling as well and provides guidance around emergency preparedness for a range travelers to include commuters, *Commuter Emergency Plan* and holiday travelers. In particular, Ready.gov has a number of emergency preparations specifically designed for people with cars. These preparations include having a full gas tank, maintaining adequate windshield wiper fluid and other basic car maintenance tips. Items specific to an emergency include:

- Jumper cables: might want to include flares or reflective triangle

- Flashlights, to include extra batteries (batteries have less power in colder weather)

- First Aid Kit, to include any necessary medications, baby formula and diapers if caring for small children

- Non-perishable food such as canned food (be alert to liquids freezing in colder weather), and protein rich foods like nuts and energy bars

- Manual can opener

- At least 1 gallon of water per person a day for at least 3 days (be alert to hazards of frozen water and resultant container rupture)

- Basic toolkit: pliers, wrench, screwdriver

- Pet supplies: food and water

- Radio: battery or hand cranked

- For snowy areas: cat litter or sand for better tire traction; shovel; ice scraper; warm clothes, gloves, hat, sturdy boots, jacket and an extra change of clothes

- Blankets or sleeping bags

- Charged Cell Phone: and car charger

In addition to emergency supplies and training for various situations, FEMA offers advice on how to mitigate disasters. The Agency gives instructions on how to retrofit a home to minimize hazards from a Flood, to include installing a Backflow prevention device, anchoring fuel tanks and relocating electrical panels.

Marked gas shuttoff

Given the explosive danger posed by natural gas leaks, Ready.gov states unequivocally that "It is vital that all household members know how to shut off natural gas" and that property owners must ensure they have any special tools needed for their particular gas hookups. Ready.gov also notes that "It is wise to teach all responsible household members where and how to shut off the electricity," cautioning that individual circuits should be shut off before the main circuit. Ready.gov further states that "It is vital that all household members learn how to shut off the water at the main house valve" and cautions that the possibility that rusty valves might require replacement.

Response

The response phase of an emergency may commence with Search and Rescue but in all cases the focus will quickly turn to fulfilling the basic humanitarian needs of the affected population. This assistance may be provided by national or international agencies and organizations. Effective coordination of disaster assistance is often crucial, particularly when many organizations respond and local emergency management agency (LEMA) capacity has been exceeded by the demand or diminished by the disaster itself. The National Response Framework is a United States government publication that explains responsibilities and expectations of government officials at the local, state, federal, and tribal levels. It provides guidance on Emergency Support Functions that may be integrated in whole or parts to aid in the response and recovery process.

On a personal level the response can take the shape either of a *shelter in place* or an *evacuation*.

In a shelter-in-place scenario, a family would be prepared to fend for themselves in their home for many days without any form of outside support. In an *evacuation*, a family leaves the area by automobile or other mode of transportation, taking with them the maximum amount of supplies they can carry, possibly including a tent for shelter. If mechanical transportation is not available, evacuation on foot would ideally include carrying at least three days of supplies and rain-tight bedding, a tarpaulin and a bedroll of blankets.

Evacuation sign

Donations are often sought during this period, especially for large disasters that overwhelm local capacity. Due to efficiencies of scale, money is often the most cost-effective donation if fraud is avoided. Money is also the most flexible, and if goods are sourced locally then transportation is minimized and the local economy is boosted. Some donors prefer to send gifts in kind, however these items can end up creating issues, rather than helping. One innovation by Occupy Sandy volunteers is to use a donation registry, where families and businesses impacted by the disaster can make specific requests, which remote donors can purchase directly via a web site.

Medical considerations will vary greatly based on the type of disaster and secondary effects. Survivors may sustain a multitude of injuries to include lacerations, burns, near drowning, or crush syndrome.

Recovery

The recovery phase starts after the immediate threat to human life has subsided. The immediate goal of the recovery phase is to bring the affected area back to normalcy as quickly as possible. During reconstruction it is recommended to consider the location or construction material of the property.

The most extreme home confinement scenarios include war, famine and severe epidemics and may last a year or more. Then recovery will take place inside the home. Planners for these events usually buy bulk foods and appropriate storage and preparation equipment, and eat the food as part of normal life. A simple balanced diet can be constructed from vitamin pills, whole-meal wheat, beans, dried milk, corn, and cooking oil. One should add vegetables, fruits, spices and meats, both prepared and fresh-gardened, when possible.

As a Profession

Professional emergency managers can focus on government and community preparedness, or private business preparedness. Training is provided by local, state, federal and private organizations and ranges from public information and media relations to high-level incident command and tactical skills.

In the past, the field of emergency management has been populated mostly by people with a military or first responder background. Currently, the field has become more diverse, with many managers coming from a variety of backgrounds other than the military or first responder fields. Educational opportunities are increasing for those seeking undergraduate and graduate degrees in emergency management or a related field. There are over 180 schools in the US with emergency management-related programs, but only one doctoral program specifically in emergency management.

Professional certifications such as Certified Emergency Manager (CEM) and Certified Business Continuity Professional (CBCP) are becoming more common as professional standards are raised throughout the field, particularly in the United States. There are also professional organizations for emergency managers, such as the National Emergency Management Association and the International Association of Emergency Managers.

Principles

In 2007, Dr. Wayne Blanchard of FEMA's Emergency Management Higher Education Project, at the direction of Dr. Cortez Lawrence, Superintendent of FEMA's Emergency Management Institute, convened a working group of emergency management practitioners and academics to consider principles of emergency management. This was the first time the principles of the discipline were to be codified. The group agreed on eight principles that will be used to guide the development of a doctrine of emergency management. Below is a summary:

1. Comprehensive – consider and take into account all hazards, all phases, all stakeholders and all impacts relevant to disasters.

2. Progressive – anticipate future disasters and take preventive and preparatory measures to build disaster-resistant and disaster-resilient communities.

3. Risk-driven – use sound risk management principles (hazard identification, risk analysis, and impact analysis) in assigning priorities and resources.

4. Integrated – ensure unity of effort among all levels of government and all elements of a community.

5. Collaborative – create and sustain broad and sincere relationships among individuals and organizations to encourage trust, advocate a team atmosphere, build consensus, and facilitate communication.

6. Coordinated – synchronize the activities of all relevant stakeholders to achieve a common purpose.

7. Flexible – use creative and innovative approaches in solving disaster challenges.

8. Professional – value a science and knowledge-based approach; based on education, training, experience, ethical practice, public stewardship and continuous improvement.

A fuller description of these principles can be found at

Tools

In recent years the continuity feature of emergency management has resulted in a new concept, Emergency Management Information Systems (EMIS). For continuity and inter-operability between emergency management stakeholders, EMIS supports an infrastructure that integrates emergency plans at all levels of government and non-government involvement for all four phases of emergencies. In the healthcare field, hospitals utilize the Hospital Incident Command System (HICS), which provides structure and organization in a clearly defined chain of command.

Disaster Response Technologies

Smart Emergency Response System (SERS) prototype was built in the SmartAmerica Challenge 2013-2014, a United States government initiative. SERS has been created by a team of nine organizations led by MathWorks. The project was featured at the White House in June 2014 and described by Todd Park (U.S. Chief Technology Officer) as an exemplary achievement.

The Smart America initiative challenges the participants to build cyber-physical systems as a glimpse of the future to save lives, create jobs, foster businesses, and improve the economy. SERS primarily saves lives. The system provides the survivors and the emergency personnel with information to locate and assist each other during a disaster. SERS allows to submit help requests to a MATLAB-based mission center connecting first responders, apps, search-and-rescue dogs, a 6-feet-tall humanoid, robots, drones, and autonomous aircraft and ground vehicles. The command and control center optimizes the available resources to serve every incoming requests and generates an action plan for the mission. The Wi-Fi network is created on the fly by the drones equipped with antennas. In addition, the autonomous rotorcrafts, planes, and ground vehicles are simulated with Simulink and visualized in a 3D environment (Google Earth) to unlock the ability to observe the operations on a mass scale.

Within Other Professions

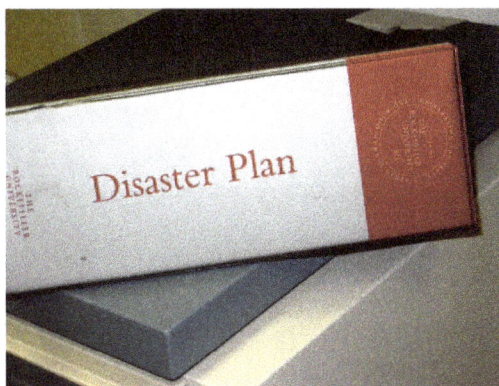

A disaster plan book at Rockefeller University in a biochemistry research laboratory.

Practitioners in emergency management come from an increasing variety of backgrounds. Professionals from memory institutions (e.g., museums, historical societies, etc.) are dedicated to preserving cultural heritage—objects and records. This has been an increasingly major component within this field as a result of the heightened awareness following the September 11 attacks in 2001, the hurricanes in 2005, and the collapse of the Cologne Archives.

To increase the potential successful recovery of valuable records, a well-established and thoroughly tested plan must be developed. This plan should emphasize simplicity in order to aid in response and recovery: employees should perform similar tasks in the response and recovery phase that they perform under normal conditions. It should also include mitigation strategies such as the installation of sprinklers within the institution. Professional associations hold regular workshops to keep individuals up to date with tools and resources in order to minimize risk and maximize recovery.

Other Tools

In 2008, the U.S. Agency for International Development created a web-based tool for estimating populations impacted by disasters. Called Population Explorer the tool uses land scan population data, developed by Oak Ridge National Laboratory, to distribute population at a resolution 1 km² for all countries in the world. Used by USAID's FEWS NET Project to estimate populations vulnerable and or imd by food insecurity, Population Explorer is gaining wide use in a range of emergency analysis and response actions, including estimating populations impacted by floods in Central America and the Pacific Ocean Tsunami event in 2009.

In 2007, a checklist for veterinarians was published in the Journal of the American Veterinary Medical Association, it had two sets of questions for a professional to ask themselves before assisting with an emergency:

Absolute Requirements for Participation:

- Have I chosen to participate?

- Have I taken ICS training?

- Have I taken other required background courses?

- Have I made arrangements with my practice to deploy?

- Have I made arrangements with my family?

Incident Participation:

- Have I been invited to participate

- Are my skill sets a match for the mission?

- Can I access just-in-time training to refresh skills or acquire needed new skills?

- Is this a self-support mission?

- Do I have supplies needed for three to five days of self-support?

While written for veterinarians, this checklist is applicable for any professional to consider before assisting with an emergency.

International Organizations

The International Emergency Management Society

The International Emergency Management Society (TIEMS), is an international non-profit NGO, registered in Belgium. TIEMS is a Global Forum for Education, Training, Certification and Policy in Emergency and Disaster Management. TIEMS' goal is to develop and bring modern emergency management tools, and techniques into practice, through the exchange of information, methodology innovations and new technologies.

TIEMS provides a platform for stakeholders to meet, network and learn about new technical and operational methodologies. TIEMS focuses on cultural differences to be understood and included in the society's events, education and research programs. This is achieved by establishing local chapters worldwide. Today, TIEMS has chapters in Benelux, Romania, Finland, Italy, Middle East and North Africa (MENA), Iraq, India, Korea, Japan and China.

International Association of Emergency Managers

The International Association of Emergency Managers (IAEM) is a non-profit educational organization aimed at promoting the goals of saving lives and property protection during emergencies. The mission of IAEM is to serve its members by providing information, networking and professional opportunities, and to advance the emergency management profession.

It has seven councils around the world: Asia, Canada, Europa, International, Oceania, Student and USA.

The Air Force Emergency Management Association, affiliated by membership with the IAEM, provides emergency management information and networking for U.S. Air Force Emergency Management personnel.

International Recovery Platform

The International Recovery Platform (IRP) was conceived at the World Conference on Disaster Reduction (WCDR) in Kobe, Hyogo, Japan in January 2005, as part of the *Hyogo Framework for Action (HFA) 2005–2015*. The HFA is a global plan for disaster risk reduction adopted by 168 governments.

The key role of IRP is to identify gaps in post disaster recovery and to serve as a catalyst for the development of tools and resources for recovery efforts.

The International Red Cross and Red Crescent Movement

The International Federation of Red Cross and Red Crescent Societies (IFRC) works closely with National Red Cross and Red Crescent societies in responding to emergencies, many times playing a pivotal role. In addition, the IFRC may deploy assessment teams, e.g. Field Assessment and Coordination Teams (FACT), to the affected country if requested by the national society. After assessing the needs, Emergency Response Units (ERUs) may be deployed to the affected country or region. They are specialized in the response component of the emergency management framework.

Baptist Global Response

Baptist Global Response (BGR) is a disaster relief and community development organization. BGR and its partners respond globally to people with critical needs worldwide, whether those needs arise from chronic conditions or acute crises such as natural disasters. While BGR is not an official entity of the Southern Baptist Convention, it is rooted in Southern Baptist life and is the international partnership of Southern Baptist Disaster Relief teams, which operate primarily in the US and Canada.

United Nations

The United Nations system rests with the Resident Coordinator within the affected country. However, in practice, the UN response will be coordinated by the UN Office for the Coordination of Humanitarian Affairs (UN-OCHA), by deploying a UN Disaster

Assessment and Coordination (UNDAC) team, in response to a request by the affected country's government.

World Bank

Since 1980, the World Bank has approved more than 500 projects related to disaster management, dealing with both disaster mitigation as well as reconstruction projects, amounting to more than US$40 billion. These projects have taken place all over the world, in countries such as Argentina, Bangladesh, Colombia, Haiti, India, Mexico, Turkey and Vietnam.

Prevention and mitigation projects include forest fire prevention measures, such as early warning measures and education campaigns; early-warning systems for hurricanes; flood prevention mechanisms (e.g. shore protection, terracing, etc.); and earthquake-prone construction. In a joint venture with Columbia University under the umbrella of the ProVention Consortium the World Bank has established a Global Risk Analysis of Natural Disaster Hotspots.

In June 2006, the World Bank, in response to the HFA, established the Global Facility for Disaster Reduction and Recovery (GFDRR), a partnership with other aid donors to reduce disaster losses. GFDRR helps developing countries fund development projects and programs that enhance local capacities for disaster prevention and emergency preparedness.

European Union

In 2001 the EU adopted Community Mechanism for Civil Protection, to facilitate co-operation in the event of major emergencies requiring urgent response actions. This also applies to situations where there may be an imminent threat as well.

The heart of the Mechanism is the Monitoring and Information Center (MIC), part of the European Commission's Directorate-General for Humanitarian Aid & Civil Protection. Accessible 24 hours a day, it gives countries access to a one-stop-shop of civil protections available amongst all the participating states. Any country inside or outside the Union affected by a major disaster can make an appeal for assistance through the MIC. It acts as a communication hub, and provides useful and updated information on the actual status of an ongoing emergency.

National Organizations

Australia

Natural disasters are part of life in Australia. Heatwaves have killed more Australians than any other type of natural disaster in the 20th century. Australia's emergency management processes embrace the concept of the prepared community. The principal government agency in achieving this is Emergency Management Australia.

Canada

Public Safety Canada is Canada's national emergency management agency. Each province is required to have both legislation for dealing with emergencies, and provincial emergency management agencies, typically called "Emergency Measures Organizations" (EMO). Public Safety Canada co-ordinates and supports the efforts of federal organizations as well as other levels of government, first responders, community groups, the private sector, and other nations. The Public Safety and Emergency Preparedness Act defines the powers, duties and functions of PS are outlined. Other acts are specific to individual fields such as corrections, law enforcement, and national security.

Germany

In Germany the Federal Government controls the German *Katastrophenschutz* (disaster relief), the Technisches Hilfswerk (*Federal Agency for Technical Relief*, THW), and the *Zivilschutz* (civil protection) programs coordinated by the *Federal Office of Civil Protection and Disaster Assistance*. Local fire department units, the German Armed Forces (Bundeswehr), the German Federal Police and the 16 state police forces (Länderpolizei) are also deployed during disaster relief operations.

There are several private organizations in Germany that also deal with emergency relief. Among these are the German Red Cross, Johanniter-Unfall-Hilfe (the German equivalent of the St. John Ambulance), the Malteser-Hilfsdienst, and the Arbeiter-Samariter-Bund. As of 2006, there is a program of study at the University of Bonn leading to the degree "Master in Disaster Prevention and Risk Governance" As a support function radio amateurs provide additional emergency communication networks with frequent trainings.

India

A protective wall built on the shore of the coastal town of Kalpakkam,
in aftermath of the 2004 Indian Ocean earthquake.

The National Disaster Management Authority is the primary government agency responsible for planning and capacity-building for disaster relief. Its emphasis is primarily on strategic risk management and mitigation, as well as developing policies and planning. The National Institute of Disaster Management is a policy think-tank and training institution for developing guidelines and training programs for mitigating disasters and managing crisis response.

The National Disaster Response Force is the government agency primarily responsible for emergency management during natural and man-made disasters, with specialized skills in search, rescue and rehabilitation. The Ministry of Science and Technology also contains an agency that brings the expertise of earth scientists and meteorologists to emergency management. The Indian Armed Forces also plays an important role in the rescue/recovery operations after disasters.

Aniruddha's Academy of Disaster Management (ACDM) is a non-profit organization in Mumbai, India with 'disaster management' as its principal objective.

Malaysia

In Malaysia, The National Security Council has the responsibility to handle emergency and disaster events. Ministry of Home Affairs Malaysia, Ministry of Health Malaysia and Ministry of Housing, Urban Wellbeing and Local Government Malaysia are also having responsibility in managing emergency. Several agencies are involved in emergency managements are Royal Malaysian Police, Malaysian Fire and Rescue Department, Malaysian Civil Defence Force, Ministry of Health Malaysia and Malaysian Maritime Enforcement Agency. There were also some voluntary organisation who involved themselves in emergency/ disaster management such as St. John Ambulance of Malaysia, Malaysian Red Crescent Society and so on.

New Zealand

In New Zealand, depending on the scope of the emergency/disaster, responsibility may be handled at either the local or national level. Within each region, local governments are organized into 16 Civil Defence Emergency Management Groups (CMGs). If local arrangements are overwhelmed, pre-existing mutual-support arrangements are activated. Central government has the authority to coordinate the response through the National Crisis Management Centre (NCMC), operated by the Ministry of Civil Defence & Emergency Management (MCDEM). These structures are defined by regulation, and explained in *The Guide to the National Civil Defence Emergency Management Plan 2006*, roughly equivalent to the U.S. Federal Emergency Management Agency's National Response Framework.

New Zealand uses unique terminology for emergency management. Emergency management is rarely used, many government publications retaining the use of the term civil defence. For example, the Minister of Civil Defence is responsible for the MCDEM. Civil Defence Emergency Management is a term in its own right, defined by statute.

And disaster rarely appears in official publications, emergency and incident being the preferred terms, with the term event also being used. For example, publications refer to the Canterbury Snow Event 2002.

"4Rs" is the emergency management cycle used in New Zealand, its four phases are known as:

- Reduction = Mitigation

- Readiness = Preparedness

- Response

- Recovery

Pakistan

Disaster management in Pakistan revolves around flood disasters focusing on rescue and relief. There is a dearth of knowledge and information about hazard identification, risk assessment and management, and disaster preparedness. Disaster management, development planning and environmental management institutions operate in isolation with no integrated planning, there being no central authority for integrated disaster management. State-level measures are heavily tilted towards structural aspects.

Russia

In Russia, the Ministry of Emergency Situations (EMERCOM) is engaged in fire fighting, civil defense, and search and rescue after both natural and human-made disasters.

Somalia

In Somalia, the Federal Government announced in May 2013 that the Cabinet had approved draft legislation on a new Somali Disaster Management Agency (SDMA), which had originally been proposed by the Ministry of Interior. According to the Prime Minister's Media Office, the SDMA will lead and coordinate the government's response to various natural disasters. It is part of a broader effort by the federal authorities to re-establish national institutions. The Federal Parliament is now expected to deliberate on the proposed bill for endorsement after any amendments.

The Netherlands

In the Netherlands the Ministry of Security and Justice is responsible for emergency preparedness and emergency management on a national level and operates a national crisis centre (NCC). The country is divided into 25 safety regions (veiligheidsregio). In a safety region, there are four components: the regional fire department, the regional department for medical care(ambulances and psycho-sociological care etc.), the regional dispatch and a section for risk- and crisis management. The regional dispatch operates for police, fire department and the regional medical care. The dispatch has all these three services combined into one dispatch for the best multi-coordinated response to an incident or an emergency. And also facilitates in information management, emergency communication and care of citizens. These services are the main structure for a response to an emergency. It can happen that, for a specific emergency, the co-operation with an other service is needed, for instance the Ministry of Defence, water board(s) or Rijkswaterstaat. The veiligheidsregio can integrate these other services into their structure by adding them to specific conferences on operational or administrative level.

All regions operate according to the Coordinated Regional Incident Management system.

United Kingdom

Following the 2000 fuel protests and severe flooding that same year, as well as the foot-and-mouth crisis in 2001, the United Kingdom passed the Civil Contingencies Act 2004 (CCA). The CCA defined some organisations as Category 1 and 2 Responders, setting responsibilities regarding emergency preparedness and response. It is managed by the Civil Contingencies Secretariat through Regional Resilience Forums and local authorities.

Disaster Management training is generally conducted at the local level, and consolidated through professional courses that can be taken at the Emergency Planning College. Diplomas, undergraduate and postgraduate qualifications can be gained at universities throughout the country. The Institute of Emergency Management is a charity, established in 1996, providing consulting services for the government, media and commercial sectors. There are a number of professional societies for Emergency Planners including the Emergency Planning Society and the Institute of Civil Protection and Emergency Management.

One of the largest emergency exercises in the UK was carried out on 20 May 2007 near Belfast, Northern Ireland: a simulated plane crash-landing at Belfast International Airport. Staff from five hospitals and three airports participated in the drill, and almost 150 international observers assessed its effectiveness.

United States

Disaster management in the United States has utilized the functional All-Hazards approach for over 20 years, in which managers develop processes (such as communication & warning or sheltering) rather than developing single-hazard or threat focused plans (e.g., a tornado plan). Processes are then mapped to specific hazards or threats, with the manager looking for gaps, overlaps, and conflicts between processes.

Given these notions, emergency managers must identify, contemplate, and assess possible man-made threats and natural threats that may affect their respective locales. Because of geographical differences throughout the nation, a variety of different threats affect communities among the states. Thus, although similarities may exist, no two emergency plans will be completely identical. Additionally, each locale has different resources and capacities (e.g., budgets, personnel, equipment, etc.) for dealing with emergencies. Each individual community must craft its own unique emergency plan that addresses potential threats that are specific to the locality.

This creates a plan more resilient to unique events because all common processes are defined, and encourages planning done by the stakeholders who are closer to the individual processes, such as a traffic management plan written by public works director. This type of planning can lead to conflict with non-emergency management regulatory bodies, which require development of hazard/threat specific plans, such as development of specific H1N1 flu plans and terrorism-specific plans.

In the United States, all disasters are initially local, with local authorities, with usually a police, fire, or EMS agency, taking charge. Many local municipalities may also have a separate dedicated office of emergency management (OEM), along with personnel and equipment. If the event becomes overwhelming to local government, state emergency management (the primary government structure of the United States) becomes the controlling emergency management agency. Federal Emergency Management Agency (FEMA), part of the Department of Homeland Security (DHS), is lead federal agency for emergency management. The United States and its territories are broken down into ten regions for FEMA's emergency management purposes. FEMA supports, but does not override, state authority.

The Citizen Corps is an organization of volunteer service programs, administered locally and coordinated nationally by DHS, which seek to mitigate disasters and prepare the population for emergency response through public education, training, and outreach. Most disaster response is carried out by volunteer organizations. In the US, the Red Cross is chartered by Congress to coordinate disaster response services, including typically being the lead agency handling shelter and feeding of evacuees. Religious organizations, with their ability to provide volunteers quickly, are usually integral during the response process. The largest being the Salvation Army, with a primary focus on chaplaincy and rebuilding, and Southern Baptists who focus on food preparation and distribution, as well as cleaning up after floods and fires, chaplaincy, mobile shower units, chainsaw crews and more. With over 65,000 trained volunteers Southern Baptist Disaster Relief is one of the largest disaster relief organizations in the US. Similar services are also provided by Methodist Relief Services, the Lutherans, and Samaritan's Purse. Unaffiliated volunteers show up at most large disasters. To prevent abuse by criminals and for the safety of the volunteers, procedures have been implemented within most response agencies to manage and effectively use these 'SUVs' (Spontaneous Unaffiliated Volunteers).

The US Congress established the Center for Excellence in Disaster Management and Humanitarian Assistance (COE) as the principal agency to promote disaster preparedness in the Asia-Pacific region.

The National Tribal Emergency Management Council (NEMC) is a non-profit educational organization developed for Tribal organizations to share information and best practices, as well as discussing issues regarding public health and safety, emergency management and homeland security, affecting those under Indian sovereignty. NTMC is organized into Regions, based on the FEMA 10 region system. NTMC was founded by the Northwest Tribal Emergency Management Council (NWTEMC), a consortium of 29 Tribal Nations and Villages in Washington, Idaho, Oregon and Alaska.

If a disaster or emergency is declared to be terror related or an "Incident of National Significance", the Secretary of Homeland Security will initiate the National Response Framework (NRF). The NRF allows the integration of federal resources with local, country, state, or tribal entities, with management of those resources to be handled at the lowest possible level, utilizing the National Incident Management System (NIMS).

FEMA's Emergency Management Institute

Emergency Management Institute's Main Campus in Emmitsburg, Maryland

The Emergency Management Institute (EMI) serves as the national focal point for the development and delivery of emergency management training to enhance the capabilities of state, territorial, local, and tribal government officials; volunteer organizations; FEMA's disaster workforce; other Federal agencies; and the public and private sectors to minimize the impact of disasters and emergencies on the American public. EMI curricula are structured to meet the needs of this diverse audience with an emphasis on separate organizations working together in all-hazards emergencies to save lives and protect property. Particular emphasis is placed on governing doctrine such as the National Response Framework (NRF), National Incident Management System (NIMS), and the National Preparedness Guidelines. EMI is fully accredited by the International Association for Continuing Education and Training (IACET) and the American Council on Education (ACE).

Approximately 5,500 participants attend resident courses each year while 100,000 individuals participate in non-resident programs sponsored by EMI and conducted by state emergency management agencies under cooperative agreements with FEMA. Another 150,000 individuals participate in EMI-supported exercises, and approximately 1,000 individuals participate in the Chemical Stockpile Emergency Preparedness Program (CSEPP).

The *independent study* program at EMI consists of free courses offered to United States citizens in Comprehensive Emergency Management techniques. Course IS-1 is entitled "Emergency Manager: An Orientation to the Position" and provides background information on FEMA and the role of emergency managers in agency and volunteer organization coordination. The EMI Independent Study (IS) Program, a Web-based distance learning program open to the public, delivered extensive online training with approximately 200 courses and trained more than 2.8 million individuals. The EMI IS Web site receives 2.5 to 3 million visitors a day.

Peak Ground Acceleration

Peak ground acceleration (PGA) is equal to the maximum ground acceleration that occurred during earthquake shaking at a location. PGA is equal to the amplitude of the largest absolute acceleration recorded on an accelerogram at a site during a particular earthquake. Earthquake shaking generally occurs in all three directions. Therefore, PGA is often split into the horizontal and vertical components. Horizontal PGAs are generally larger than those in the vertical direction but this is not always true, especially close to large earthquakes. PGA is an important parameter (also known as an intensity measure) for earthquake engineering, The design basis earthquake ground motion (DBEGM) is often defined in terms of PGA.

Unlike the Richter and moment magnitude scales, it is not a measure of the total energy (magnitude, or size) of an earthquake, but rather of how hard the earth shakes at a given geographic point. The Mercalli intensity scale uses personal reports and observations to measure earthquake intensity but PGA is measured by instruments, such as accelerographs. It can be correlated to macroseismic intensities on the Mercalli scale but these correlations are associated with large uncertainty.

The peak horizontal acceleration (PHA) is the most commonly used type of ground acceleration in engineering applications. It is often used within earthquake engineering (including seismic building codes) and it is commonly plotted on seismic hazard maps. In an earthquake, damage to buildings and infrastructure is related more closely to ground motion, of which PGA is a measure, rather than the magnitude of the earthquake itself. For moderate earthquakes, PGA is a reasonably good determinant of damage; in severe earthquakes, damage is more often correlated with peak ground velocity.

Geophysics

Earthquake energy is dispersed in waves from the hypocentre, causing ground movement omnidirectionally but typically modelled horizontally (in two directions) and vertically. PGA records the acceleration (rate of change of speed) of these movements, while peak ground velocity is the greatest speed (rate of movement) reached by the ground, and peak displacement is the distance moved. These values vary in different earthquakes, and in differing sites within one earthquake event, depending on a number of factors. These include the length of the fault, magnitude, the depth of the quake, the distance from the epicentre, the duration (length of the shake cycle), and the geology of the ground (subsurface). Shallow-focused earthquakes generate stronger shaking (acceleration) than intermediate and deep quakes, since the energy is released closer to the surface.

Peak ground acceleration can be expressed in g (the acceleration due to Earth's gravity, equivalent to g-force) as either a decimal or percentage; in m/s² ($1\,g$ = 9.81 m/s²); or in Gal, where 1 Gal is equal to 0.01 m/s² ($1\,g$ = 981 Gal).

The ground type can significantly influence ground acceleration, so PGA values can display extreme variability over distances of a few kilometers, particularly with moderate to large earthquakes. The varying PGA results from an earthquake can be displayed on a shake map. Due to the complex conditions affecting PGA, earthquakes of similar magnitude can offer disparate results, with many moderate magnitude earthquakes generating significantly larger PGA values than larger magnitude quakes.

During an earthquake, ground acceleration is measured in three directions: vertically (V or UD, for up-down) and two perpendicular horizontal directions (H1 and H2), often north-south (NS) and east-west (EW). The peak acceleration in each of these directions is recorded, with the highest individual value often reported. Alternatively, a combined value for a given station can be noted. The peak horizontal ground acceleration (PHA or PHGA) can be reached by selecting the higher individual recording, taking the mean of the two values, or calculating a vector sum of the two components. A three-component value can also be reached, by taking the vertical component into consideration also.

In seismic engineering, the effective peak acceleration (EPA, the maximum ground acceleration to which a building responds) is often used, which tends to be ⅔ – ¾ the PGA.

Seismic Risk and Engineering

Study of geographic areas combined with an assessment of historical earthquakes allows geologists to determine seismic risk and to create seismic hazard maps, which show the likely PGA values to be experienced in a region during an earthquake, with a probability of exceedance (PE). Seismic engineers and government planning departments use these values to determine the appropriate earthquake loading for buildings in each zone, with key identified structures (such as hospitals, bridges, power plants) needing to survive the maximum considered earthquake (MCE).

Damage to buildings is related to both peak ground velocity and PGA, and the duration of the earthquake – the longer high-level shaking persists, the greater the likelihood of damage.

Comparison of Instrumental and Felt Intensity

Peak ground acceleration provides a measurement of *instrumental intensity*, that is, ground shaking recorded by seismic instruments. Other intensity scales measure *felt intensity*, based on eyewitness reports, felt shaking, and observed damage. There is correlation between these scales, but not always absolute agreement since experiences and damage can be affected by many other factors, including the quality of earthquake engineering.

Generally Speaking,

- 0.001 g (0.01 m/s^2) – perceptible by people

- 0.02 g (0.2 m/s^2) – people lose their balance

- 0.50 g – very high; well-designed buildings can survive if the duration is short.

Correlation with the Mercalli Scale

The United States Geological Survey developed an Instrumental Intensity scale which maps peak ground acceleration and peak ground velocity on an intensity scale similar to the felt Mercalli scale. These values are used to create shake maps by seismologists around the world.

Instrumental Intensity	Acceleration (g)	Velocity (cm/s)	Perceived Shaking	Potential Damage
I	< 0.0017	< 0.1	Not felt	None
II-III	0.0017 - 0.014	0.1 - 1.1	Weak	None
IV	0.014 - 0.039	1.1 - 3.4	Light	None
V	0.039 - 0.092	3.4 - 8.1	Moderate	Very light
VI	0.092 - 0.18	8.1 - 16	Strong	Light
VII	0.18 - 0.34	16 - 31	Very strong	Moderate
VIII	0.34 - 0.65	31 - 60	Severe	Moderate to heavy
IX	0.65 - 1.24	60 - 116	Violent	Heavy
X+	> 1.24	> 116	Extreme	Very heavy

Other Intensity Scales

In the 7-class Japan Meteorological Agency seismic intensity scale, the highest intensity, Shindo 7, covers accelerations greater than 4 m/s² (0.41 *g*).

PGA Hazard Risks Worldwide

In India, areas with expected PGA values higher than 0.36*g* are classed as "Zone 5", or "Very High Damage Risk Zone".

Notable Earthquakes

PGA single direction (max recorded)	PGA vector sum (H1, H2, V) (max recorded)	Mag	Depth	Fatalities	Earthquake
2.7g	2.99 g	9.0	30 km	>15,000	2011 Tōhoku earthquake and tsunami
2.2g		6.3	5 km	185	February 2011 Christchurch earthquake
2.13g		6.4	6 km	1	June 2011 Christchurch earthquake
	4.36g	6.9/7.2	8 km	12	2008 Iwate-Miyagi Nairiku earthquake
1.7g		6.7	19 km	57	1994 Los Angeles earthquake
	1.47g	7.1	42 km	4	April 2011 Miyagi earthquake
1.26g		7.1	10 km	0	2010 Canterbury earthquake
1.01g		6.6	10 km	11	2007 Chūetsu offshore earthquake
1.01g		7.3	8 km	2,415	1999 Jiji earthquake
1.0g		6.0	8 km	0	December 2011 Christchurch earthquake
0.8g		6.8	16 km	6,434	1995 Kobe earthquake
0.78g		8.8	23 km	521	2010 Chile earthquake
0.6g		6.0	10 km	143	1999 Athens earthquake
0.51g		6.4		612	2005 Zarand earthquake
0.5g		7.0	13 km	92,000-316,000	2010 Haiti earthquake
0.438g		7.7	44 km	27	1978 Miyagi earthquake (Sendai)
0.4g		5.7	8 km	0	2016 Christchurch earthquake
0.367g		5.2	1 km	9	2011 Lorca earthquake
0.25 - 0.3g		9.5	33 km	1,655	1960 Valdivia earthquake
0.24g		6.4		628	2004 Morocco earthquake
0.18g		9.2	23 km	143	1964 Alaska earthquake

References

- Buchanan, Sally. "Emergency preparedness." from Paul Banks and Roberta Pilette. Preservation Issues and Planning. Chicago: American Library Association, 2000. 159–165. ISBN 978-0-8389-0776-4

- Doss, Daniel; Glover, William; Goza, Rebecca; Wigginton, Michael (2015). The Foundations of Communication in Criminal Justice Systems (1 ed.). Boca Raton, Florida: CRC Press. p. 301. ISBN 978-1482236576.

- Slyke, Jeffrey; Nations, Julie (2014). Foundations of Emergency Management (1st ed.). Dubuque, IA: Kendall-Hunt Publishing Company. p. 25. ISBN 978-1465234889.

- McElreath, David; Doss, Daniel; Jensen, Carl; Wigginton, Michael; Nations, Robert; Van "The Salvation Army Emergency Disaster Services". Disaster.salvationarmyusa.org. Retrieved 2015-03-08.

- European Facilities for Earthquake Hazard & Risk (2013). "The 2013 European Seismic Hazard Model (ESHM13)". EFEHR. Retrieved 2015-11-11.

- "Masuring Progress in Chemical Safety : A Guide for Local Emergency Planning Committees and Similar Groups" (PDF). Epa.gov. Retrieved 2015-03-08.

- Baird, Malcolm E. (2010). "The "Phases" of Emergency Management" (PDF). Vanderbilt Center for Transportation Research. Retrieved 2015-03-08.

- "School-Based Emergency preparedness : A National analysis and recommended Protocol" (PDF). Archive.ahrq.gov. Retrieved 2015-03-08.

- "Preparedness 101: Zombie Apocalypse | Public Health Matters Blog | Blogs | CDC". Blogs.cdc.gov. 2011-05-16. Retrieved 2015-03-08.

- "CDC Emergency preparedness and You | Gather Emergency Supplies | Disaster Supplies Kit". Emergency.cdc.gov. Retrieved 2015-03-08.

- "Preparing for Disaster for People with Disabilities and other Special Needs" (PDF). Redcross.org. Retrieved 2015-03-08.

- "The Smart Emergency Response System Using MATLAB and Simulink". YouTube. 2014-07-18. Retrieved 2015-03-08.

- "Welcome to the International Recovery Platform — International Recovery Platform". Recovery-platform.org. Retrieved 2015-01-15.

- "Institute of Civil Protection & Emergency Management | Welcome". ICPEM. 2014-04-03. Retrieved 2015-03-08.

- "Duromax RV Grade 4,400-Watt 7.0 HP Gasoline Powered Portable Generator with Wheel Kit-XP4400 - The Home Depot". The Home Depot. 15 October 2014. Retrieved 8 March 2015.

- "Civil Defence Emergency Management Act 2002 No 33 (as at 01 January 2014), Public Act 4 Interpretation – New Zealand Legislation". Legislation.govt.nz. Retrieved 2015-03-08.

- "Maine Emergency Management Agency" (2007). "What is Emergency Management?". Retrieved 2014-02-22.

- NEWS: Pakistan's Punjab builds model villages to withstand disasters, Climate & Development Knowledge Network, 17 December 2013.

- AHRQ Publication No. 09-0013 January 2009 UMass System Office Hazard Mitigation Plan Draft (December 2013)

Permissions

All chapters in this book are published with permission under the Creative Commons Attribution Share Alike License or equivalent. Every chapter published in this book has been scrutinized by our experts. Their significance has been extensively debated. The topics covered herein carry significant information for a comprehensive understanding. They may even be implemented as practical applications or may be referred to as a beginning point for further studies.

We would like to thank the editorial team for lending their expertise to make the book truly unique. They have played a crucial role in the development of this book. Without their invaluable contributions this book wouldn't have been possible. They have made vital efforts to compile up to date information on the varied aspects of this subject to make this book a valuable addition to the collection of many professionals and students.

This book was conceptualized with the vision of imparting up-to-date and integrated information in this field. To ensure the same, a matchless editorial board was set up. Every individual on the board went through rigorous rounds of assessment to prove their worth. After which they invested a large part of their time researching and compiling the most relevant data for our readers.

The editorial board has been involved in producing this book since its inception. They have spent rigorous hours researching and exploring the diverse topics which have resulted in the successful publishing of this book. They have passed on their knowledge of decades through this book. To expedite this challenging task, the publisher supported the team at every step. A small team of assistant editors was also appointed to further simplify the editing procedure and attain best results for the readers.

Apart from the editorial board, the designing team has also invested a significant amount of their time in understanding the subject and creating the most relevant covers. They scrutinized every image to scout for the most suitable representation of the subject and create an appropriate cover for the book.

The publishing team has been an ardent support to the editorial, designing and production team. Their endless efforts to recruit the best for this project, has resulted in the accomplishment of this book. They are a veteran in the field of academics and their pool of knowledge is as vast as their experience in printing. Their expertise and guidance has proved useful at every step. Their uncompromising quality standards have made this book an exceptional effort. Their encouragement from time to time has been an inspiration for everyone.

The publisher and the editorial board hope that this book will prove to be a valuable piece of knowledge for students, practitioners and scholars across the globe.

Index

Chinese Medicine: A Comprehensive Guide